Political Mobilisation in India

This book addresses the paradox of political mobilisation and the failings of governance in India, with reference to the conflict between secularism and Hindu nationalism, authoritarianism and democracy. It demonstrates how the Internal Emergency of 1975 led to increased support of groups such as the BJS, which by 1980 became the BJP, and the RSS accounting for the rise of political movements advocating Hindu nationalism – *Hindutva* – as a response to rapid political mobilisation triggered by the Emergency, and an attempt by political elites to control this to their advantage. It argues that the political disjuncture between democracy and mobilisation in India is partly a function of the Indian state, the nature of a caste-class based society, but also – and significantly – the contingencies of individual leaders and the styles of rule. It shows how, in the wake of the Emergency, the BJP and the RSS gained popularity and power amid the on-going decline and fragmentation of the Congress, whilst, at the same time, Hindu nationalism appeared to be of such importance that Congress began aligning themselves with the Hindu right for electoral gains. It suggests that, in the light of these developments, the rise of the BJP should not be considered as remarkable – or as transformative – as was at first imagined. By relocating recent debates about secularism and Hindu nationalism within a broader historical framework of Congress nationalism and Congress decline, this book is a substantive and original contribution to the study of Indian politics.

Vernon Hewitt is Senior Lecturer in the Department of Politics at the University of Bristol, UK. He is the author of *The International Politics of South Asia* (1992), *Reclaiming the Past: Jammu and Kashmir* (1996) and numerous articles and chapters on ethnicity and the post-colonial state, development and colonial history.

Routledge Advances in South Asian Studies
Edited by Subrata K. Mitra
South Asia Institute, University of Heidelberg, Germany

South Asia, with its burgeoning, ethnically diverse population, soaring economies, and nuclear weapons, is an increasingly important region in the global context. The series, which builds on this complex, dynamic and volatile area, features innovative and original research on the region as a whole or on the countries. Its scope extends to scholarly works drawing on history, politics, development studies, sociology and economics of individual countries from the region as well as those that take an interdisciplinary and comparative approach to the area as a whole or to a comparison of two or more countries from this region. In terms of theory and method, rather than basing itself on any one orthodoxy, the series draws broadly on the insights germane to area studies, as well as the tool kit of the social sciences in general, emphasising comparison, the analysis of the structure and processes, and the application of qualitative and quantitative methods. The series welcomes submissions from established authors in the field as well as from young authors who have recently completed their doctoral dissertations.

1 **Perception, Politics and Security in South Asia**
The compound crisis of 1990
P. R. Chari, Pervaiz Iqbal Cheema and Stephen Philip Cohen

2 **Coalition Politics and Hindu Nationalism**
Edited by Katharine Adeney and Lawrence Saez

3 **The Puzzle of India's Governance**
Culture, context and comparative theory
Subrata K. Mitra

4 **India's Nuclear Bomb and National Security**
Karsten Frey

5 **Starvation and India's Democracy**
Dan Banik

6 **Parliamentary Control and Government Accountability in South Asia**
A comparative analysis of Bangladesh, India and Sri Lanka
Taiabur Rahman

7 **Political Mobilisation and Democracy in India**
States of emergency
Vernon Hewitt

Political Mobilisation and Democracy in India

States of emergency

Vernon Hewitt

Routledge
Taylor & Francis Group

LONDON AND NEW YORK

First published 2008
by Routledge
2 Park Square, Milton Park, Abingdon, Oxon OX14 4RN

Simultaneously published in the USA and Canada
by Routledge
270 Madison Ave, New York, NY 10016

Routledge is an imprint of the Taylor & Francis Group, an informa business

Transferred to Digital Printing 2009

© 2008 Vernon Hewitt

Typeset in Times New Roman by
Newgen Imaging Systems (P) Ltd, Chennai, India

British Library Cataloguing in Publication Data
A catalogue record for this book is available from the British Library

Library of Congress Cataloging in Publication Data
A catalog record for this book has been requested

ISBN10: 0–415–42375–9 (hbk)
ISBN10: 0–415–54479–3 (pbk)
ISBN10: 0–203–94496–8 (ebk)

ISBN13: 978–0–415–42375–5 (hbk)
ISBN13: 978–0–415–54479–5 (pbk)
ISBN13: 978–0–203–94496–7 (ebk)

To Gavin Williams
Slow train coming

The tyrant's device:
Whatever is possible is necessary

W. H. Auden.

Contents

Acknowledgements   viii
Map   ix

Introduction   1

1   Emergencies, states and societies: the study of Indian politics   13

2   State–society relations 1947–1950: a democratic polity?   40

3   Political mobilisations 1963–1971   62

4   The state and political crisis 1971–1975   93

5   The state as political impasse 1975–1980   121

6   Political mobilisations 1980–1986   150

7   *Hindutva* as crisis 1986–2004   181

Conclusions   196

Notes   206
References   210
Index   225

Acknowledgements

Many people have assisted me in the writing of this book. I would like to thank John Harriss and Gavin Williams for their intellectual encouragement – starting in a far off world without the internet, without emails and without mobile phones. I am indebted to Subrata Mitra for his encouragement and support over the years – much water has flowed under the proverbial bridge since we first met in Hull, but he has been a continuous presence for me. Andrew Wyatt, my colleague, has also been a constant source of support, putting up with my idiosyncratic work habits with good grace. I am indebted to Dan Blank of the governance research centre for his assistance in cleaning up the text, and I am particularly (or is that peculiarly) in debt to Professor Anthony Forster, now of Durham, for adopting a subtle (and not so subtle) array of 'reverse' psychological techniques to encourage me to finish this text. They worked very effectively.

There are some astounding scholars writing on South Asia, especially India. I have benefited from all of these, especially Barbara Harriss-White's book *India Working*. I have also benefited from the joy of teaching at Bristol, at both the graduate and the undergraduate level, and from being challenged by students eager to travel to and see South Asia. Long may universities commit themselves to utilising individual research both as a contribution to knowledge, and as a tool through which to achieve (and maintain) excellence in teaching.

Vernon Hewitt
Bristol
June 2007

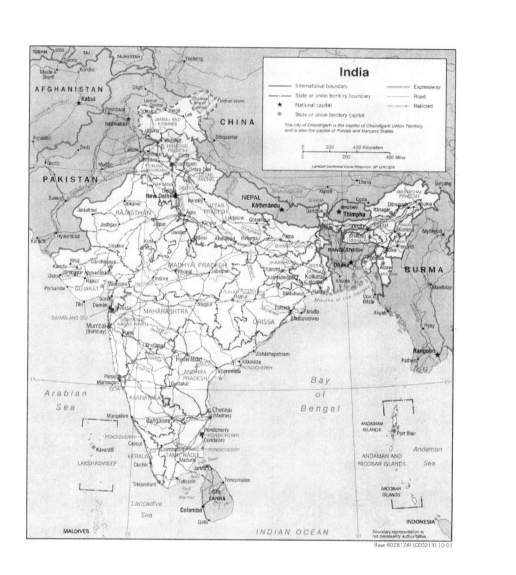

India

— International boundary
★ National capital
⊛ State or union territory capital

▬▬▬ Expressway
‥‥‥ Road
—‥— Railroad

The city of Chandigarh is the capital of Chandigarh Union Territory
and is also the capital of Punjab and Haryana States.

Lambert Conformal Conic Projection, SP 12°N | 32°N

Base 802812AI (C00213) 10-01

Boundary representation is
not necessarily authoritative.

Introduction

This book consists of a re-worked version of my doctoral thesis entitled *Locating the Internal Emergency 1967–1977*, and later research interests on Indian secularism and the Bharatiya Janata Party (BJP), undertaken in the late 1990s. As such, this work brings together two apparently diverse literatures on India: deinstitutionalisation and political centralisation, with debates on the nature and scope of secularism and *Hindutva* (a term first used in 1923 to describe movements advocating Hindu nationalism). It brings them together in the conviction that they are in fact wider parts of the same dynamics, namely the prolific mobilisation of Indian society since the late 1960s, especially 1977, and then onwards through the 1990s; and the changes that this has brought about regarding the nature of the Indian state.

The book also seeks to address the manipulative powers of political leaders in seeking to ensure their re-election and looks at how, situated within the state, they have sought to shape and direct social mobilisation. Often the results have been both catastrophic and, for what it matters, largely unintentional.[1] Locating the Emergency within the wider processes at work throughout India since 1980 adds not just to our understanding of those extraordinary years in the mid 1970s, but also accounts for what happened afterwards in terms of the inexorable rise of non-Congress governments in the states, non-Congress coalitions at the centre, and above all, the rise of, and challenge to, Hindu nationalism.

What is the nature of this link? To cite Hansen, 'the phenomenal growth and political success of Hindu nationalism in India...must be understood in the context of the larger disjuncture between democratic mobilisation and democratic governance' (Hansen 1999: 17). It will be argued here that this disjuncture was made possible because of the specific institutional and electoral peculiarities of India, and was to be widened considerably by the Emergency itself. The Emergency – and its demise – was borne out of a long-term disjuncture between ongoing democratic mobilisation from the time of independence, and the *incapacity* of democratic governance by the Indian state led by a particular party and leadership. Afterwards, in the decade that followed, the disjuncture continued apace and provided the peculiar and highly contingent paradox in which, amid the increased participation of the poor and the ongoing fragmentation of political authority, an ideological project could come to the fore that was elitist, intolerant and anti-poor.

The Emergency stands out as a critical moment in which elite-managed democratic politics broke down both within the Congress Party, and within the Indian political system generally, amid widespread mobilisation and the 'deepening' politicisation of India's population. One of the central arguments of this book is that it is within the context of the rapid incursion of 'subaltern' or 'plebeian' politics into the state, in the wake of the Emergency, that the instrumental use of Hindu anxieties becomes *nationally* legitimated, with all the consequences for social disorder, and the challenge to what has become known, with a certain amount of nostalgia, as the Nehruvian project (Adeney and Saez 2005). It is also one of the central arguments of this book that this very same context both limits and undermines *Hindutva* itself, both with regard to its problematic organisational division of labour between state and society, and to its programmatic weakness once in power.

The use of socialism, of *garibi hatao* (Indira Gandhi's populist slogan translated as 'out with poverty') and of *Hindutva* are in the first instance conceptualised as differing state strategies of co-optation, deployed by elites to electorally stem growing fragmentation and to legitimate power in the context of rapid democratic mobilisation. They are also conceived as involving different 'ideas of India', differing ways in which a nation, a collectivity, can be imagined and represented. It will be argued that while both these strategies were present at the time of independence, one was in an implicit and marginal form, partly constrained by Nehru, but for the most part by its own programmatic and policy weaknesses. Given the overall dynamics of state–society relations in India, and the contingencies of political leaders to both court and control widening participation without delivering policy, one strategy (socialism) increasingly gave way to another (*Hindutva*), to become the dominant political discourse of the 1990s, but one that was peculiarly stigmatic and which could, in electoral terms, appeal only to a narrow elite, momentarily corralled by a fragmented and divided poor.

The socio-economic agenda of *Hindutva* is confused and at times, contradictory. It speaks a language of social uplift and 'development', but also seeks to free the state and a particular social elite from the burden of the 'continuous demands, continuous noise' of the dispossessed and the disadvantaged (Kothari 1996: 432). It speaks of an authentic, national identity but remains as much a construct of the modern world as Nehru's socialism, and largely happy with the same state format as the one Nehru bequeathed. Amid an increase in socio-economic violence, it uses the language of secularism and pragmatic politics. Grounded in societal organisations, it is nonetheless profoundly political and state orientated, and surprisingly, given its relative 'electoral' weakness, the state *has repeatedly given way to it* (Bates 2005). How do we explain this large disjuncture, this paradox of political mobilisation and the faltering of democratic governance?

The argument here is that the link between *garibi hatao*, the Emergency and *Hindutva* is in part causal, but it requires the understanding of a series of contingent events: how and in what ways do political elites perceive threats to their position, and how and in what ways do they seek to neutralise them? To understand this, it is necessary to set out a particular recognition that the term state and society are closely nested concepts, indeed almost incapable of being separated out as a particular strategy unfolds. Although this book offers an explanation of the Internal Emergency of 1975 (an event which still oddly lacks an authoritative text), it also offers an explanation as to how, in the wake of the 1977 elections, the outlines of a new strategy suggested itself to Mrs Gandhi amid the disintegration of India's first non-Congress government. This strategy was premised on the bankruptcy of her previous policies, and upon the recognition of societal forces that she needed to accommodate or control, forces that had crystallised in opposition to her continuation in office but also forces that, ironically, she had released by her own political style of leadership.

In particular, the Emergency was critical in mobilising political support around the Bharatiya Jana Sangh (BJS) and the Rashtriya Swayamsevak Sangh (RSS), as key components of the Jayaprakash Narayan (JP) Movement that had organised the Bihar and Gujarat agitations from 1974. In the context of rampant corruption, and amid the belief that institutional politics was unfairly dominated and manipulated by Congress, these forces went onto resist Mrs Gandhi's attempts to retain office in the wake of the Allahabad court case by extra-constitutional means, using the experience of the RSS and the idiom of resistance to 'unjust' governments amplified by the Gandhian tradition within Congress itself. Following the declaration of the Emergency, tactics of societal resistance were used, such as the courting of mass arrests, through which the state singled out the RSS in particular as constituting a major threat to governance (Jaffrelot 1996: 275). Through repression and arrest, social activists and volunteers associated with the BJS become widely legitimated as defenders of democratic government. In shutting down the formal democratic space to opposition parties within the states, and in repressing where possible the politics of the street, Indira Gandhi's repressive measures perversely favoured the Sangathanist, non-political networks of the RSS, with their long-term emphasis on culture and identity, and strategy of penetrating and transforming Hindu society at large. The JP Movement enabled RSS and BJS activists to mobilise concerns around issues that fitted into their wider ideological concerns, but also to challenge and, through the BJS merger into the Janata government, to gain access to the state.

Post-Emergency, following the spectacular collapse of the regime, and the controversies over the associations between the RSS, the BJS and the Janata government, political elites within the Congress felt compelled to confront a wave of radical and deepening mobilisation that they had themselves triggered, playing into concerns of insecurity and violence. In the wake of Mrs Gandhi's assassination, Rajiv Gandhi moved towards a technique of societal accommodation that competitively bought off communities and weakened the state's ability to enforce a particular political discourse on secularism and regulate the symbolism of democratic competition. Already weakened by the Emergency, and made possible by the contradictory nature of secularism as practiced by the Indian central state, Rajiv Gandhi's electoral strategy, and that of the V. P. Singh government that followed, made possible the 'impossible assimilation' (Jaffrelot 1996: 283–91) between the societal projects of the RSS, the electoral logic of a political party and the democratic capturing of state power. But it could only be sustained by crisis, the structural oddities of India's electoral system and the weakness of political leadership.

Chapter 1 situates my understanding of state–society relations in the context of a broad – almost unmanageably large – literature that has grown up around the study of the Emergency itself, but which has rarely and often not systematically been applied to the study of *Hindutva*. In Chapter 1, I outline and, in places, re-emphasise some of the works of Paul Brass, Rajni Kothari, James Manor, Morris-Jones, Suzanne and Lloyd Rudolph, Atul Kohli, Sudipta Kaviraj, Pranad Bardham and especially Partha Chatterjee's recent work on governmentality. In summarising such a range of authors I have where necessary removed them from the primacy of the Emergency as an event, and reiterated their insights with reference to the ongoing democratic mobilisation of the 1980s and the rise of the BJP from 1989 onwards. In doing so I have added to their remit (if I can put it that way) debates over *Hindutva* and the crisis of secularism in India as it relates to the changes in state–society interactions through the process of mass mobilisation and political manipulation.

I then set out my own preferences for conceptualising the Indian state and its wider association with society. This has resulted in a need to interrogate a Marxist-inspired literature not so much from the perspective of capitalism as a mode of production (or in more recent fashion, the perspective of 'modernity') but from the perspective of the tensions created when states seek to manage complex and transforming societies democratically, looking at the ongoing and often unintended consequences that derive from deepening democratisation within increasingly limiting (or even anachronistic) procedures and techniques of governance. I am also interested in identifying the opportunities that present themselves in the context of rapid democratisation for a specific leadership to pursue what they conceive of as societally sanctioned goals. This involves the elites situated within the state taking on issues that have emerged within society itself and which, appear to countenance some form of recognition by a leadership seeking – almost at any cost – to gain electoral credibility and legitimacy. Significantly, the inclination by an elite to take up issues situated throughout civil society may be in large part determined by the nature of intra-elite competition for prestige and power within the institutions of the state itself (Migdal 1988).

I have been particularly influenced by the writings of Dietrich Rueschemeyer and his co-authors (Rueschemeyer *et al.* 1992). I have also drawn upon some of the insights of Theda Skocpol (Skocpol 1979), the historical comparative work of Weiss and Hobson (Weiss and Hobson 1995), and above all the state-in-society model of Migdal (Migdal 1988, 2004). Such works address what are in effect a series of continuous interactions between state institutions, civil society and social classes – identified broadly and pragmatically in relationship to economic production, ownership and access to resources. The utility of such approaches is that there is an explicit emphasis on the state as a set of historically grounded institutions free of any a priori assumptions as to how they relate to societal-based interests, particularly social class. The state is not reduced to dominant class interests, but *can* be conceived as an 'actor in its own right, and the preferences, projects and policies of a state elite can be generated independently of other power sources situated unevenly within civil society' (Weiss and Hobson 1995: 10), an idea discussed elsewhere as the 'state as organisation'.[2]

Furthermore, and drawing again on Migdal, it is necessary to see the state as a multiple set of actors and institutions. Too often the state is not only conceived as captive of societal interests, but is personified as a unitary actor. It should rather be the case that the state is perceived as encompassing 'complex layers of political institutions that operate in a variety of ways in differing contexts. Indeed, instead of talking about the Indian state as a monolithic entity, it is preferable to talk about the Indian state as a sum total of myriad forms' (Bates and Basu 2005: xiii).

In Migdal's accounts, the state in society provides the site for intense mobilisation and counter-mobilisations between differing groups, and given the varied, myriad structure of the state, this can and does manifest itself as a series of contestations between differing sections of the state itself. State capacities become not only something intrinsic to the idea of the state itself, but also derive from the society to which the state is relating, what dominant interests exist at a particular level, and what consequences follow when state elites seek to co-opt them or manipulate their apparent significance. It is also informed by the particular historical evolution of state institutions, and their relationship to any party structure or system. To cite Mann:

> The state contains two dualities: it is practice and persons, and centre and territory. Political power is simultaneously 'statist'. Vested in elite persons and institutions at the centre and it is composed of 'party relations' between persons and institutions of the centre and across state

territories. Thus it will crystallise in forms essentially generated by the outside society and in forms that are intrinsic to its own political process.

(Mann 1993: 56)

The advantage of Midgal's approach is the localist, almost ethnographic way in which the state is conceived. In tandem with critical ethnographic works that look at how local politics informally configure and redefine state action, often in opposition to the formal strategies of democratisation adopted by states and international agencies, there is a clear empirical richness here. Ethnographic works also stress the importance of cultural movements and the politics of cultural mobilisation that have, especially in the case of India, organised themselves outside the more formally defined, institutional frameworks of the state (see Zavos *et al.* 2004). It does not follow, however, that they are indifferent to, or unaffected by, the state itself, or that the state is unaware of them.

Likewise, with reference to society, I have drawn on a more dynamic understanding of the term than implied by discussions of civil society and social capital set out in the more mainstream political science literature on democratic transition (Burnell 2002; Burnell and Ware 1998; Jha 1998: 63–96;). Again, this implies a rather misleading conceptual gap as to where the state ends and society begins. Chatterjee's distinction between civil and political society is used to distinguish between the small elite space of voluntary, middle class social movements, and the wider more plebeian forms of popular participation by groups long excluded from the niceties of the constitution, and which involve paralegal and criminal forms of activity (Chatterjee 2004: 27–80). Like Harriss-White's debate's on the 'shadow-state', Chatterjee's work (although in its broader, post-structural moments, deeply problematic) enables a sophisticated understanding of the state to be matched with an ethnographic understanding of how local movements and forces that emerge within political society (violent, illegal and communal) are often at odds with world civil society and legitimate constitutionalism and party politics. The resulting dynamics of these contradictions set off further crisis and compromise within political elites that fundamentally change the nature of governance.

Again, Chatterjee highlights how the state is experienced and indeed perceived at local and subaltern levels in a way that is ignored within the more mainstream literature, where often the state as defined by the democratisation debate – unified, homogenised and autonomous – is an academic fiction (see Hansen 2001; Harriss-White 2003). To cite Jeremy Gould:

The image of the...state...lives on in the social imaginary. The state of recent anthropological interest is less an efficacious regulatory force, than a quasi-mythical entity with which competing actors attempt to associate and thus legitimise their claims to public authority.

(Gould 2003: 32)

This legitimation takes place by capturing or bypassing the local or regional institutions of the state, and by manipulating and distorting its ideological agenda for one's own purposes, but state capture can likewise fragment or distort the social imaginary itself. It follows from this instance of a state-in-society model that recognising cultural or religious signifiers free from politics – and defining societal movements as separate from the state – should be undertaken with extreme caution, since they are often momentary crystallisations of power.

Having set out the relevant literature of state–society relations, democratisation and *Hindutva*, Chapter 2 sketches the historical context in which these arguments are applied to India. Although triggered by a series of contingent events, the consequences of the Emergency require a study of the wider historical context of the Indian National Congress, and the nature of state–society relations in India that stress the links between the contemporary state and the colonial state. In the context of colonialism, the Congress, the nationalist elite, the dynamics of caste and the institutions of state, crystallised power in a rather specific and initially quite localised way. Such intimacy gave the party considerable power to centralise and formally dominate, and gave the Indian state apparent autonomy over society at the time of independence, especially the central institutions of government.

Away from the formal constitution, democracy in India was a narrow affair, in which a small group of strategically placed elites delivered to the Congress large majorities to retain office through election. The logic of politics was as much about control as it was about genuine participation and freedom. Until the late 1960s, India's political elite can be adequately identified through Reuschmeyer's term as constituting 'a liberal oligarchy'. Or, to put it another way, 'the modernising, developmental mission of the new state...was in an important sense one that was imposed by an elite which was not notably democratic in its own attitudes or actions and which had a history of negating expressions of popular will' (Corbridge and Harriss 2000: xvii).

As an electoral machine, the Congress system deployed the language of a secular and socialist nationalism at the centre, while consolidating a series of dominant social castes in the countryside, and working alongside Hindu traditionalists at the local level and in North India, the state level in particular.

The Nehruvian rhetoric and idea of India was, at the federal level, a secular one committed to social emancipation and to economic growth through a mixed economy, involving the state in the management of capitalist enterprises and the reform of land ownership. Yet this modernity overlay a broader set of regionalised interests in which ideas of religion, community and culture were redefined and readily adapted to the specifics of the modern state in ways that would complicate the modernism of Nehru. For instance, the commitment to socialism and the redistribution of social assets and resources were premised on the recognition of communalised economic rights embedded within cultural and religious signifiers

(Madan 1998). Individuals were often seen as discrete members of either advantaged or disadvantaged communities, which were in turn defined through religious and cultural practices. As such, the apparently obvious distinction between culture and socio-economics rights, between religion and the state, so central to India's commitment to modernisation, was far less clear than many politicians (or, for that matter, academics) claimed. At local levels of power, it arguably disappeared altogether. In the particular case of India 'Secularism became in the post-colonial mass democracy a privileged signifier of equal accommodation and competitive patronage of social groups and cultural communities through state and party' (Hansen 1999: 136). This initially took place at the local and regional levels, but then increasingly at the national level.

Such social and cultural groups, self-identified through the languages of religion and cultural difference, would over time perceive not so much equality than aspects of unequal practice, uneven and disparate access to state institutions, and unfair access to state resources that tied culture and religion into strategies of uneven economic accumulation and privilege. They would see, through the failure of socialism, not so much secularism, but pseudo-secularism.

This unevenness was for a time disguised by what Jaffrelot has convincingly identified as the 'hegemonic position of secularism' of the central state, and the functioning of the Congress system as a dominant electoral regime that allowed specific political idioms and societal coalitions to coexist side by side, often regionally isolated between states. As it eroded, as the system 'nationalised' and as the political elites moved to directly channel and control waves of political mobilisation by co-opting symbols of religion, culture and community in the service of affirmative or emancipatory policies, this hegemonic position was easily breached, and with consequences for Indian state–society relations that were largely unforeseen. Intense electoral competition drove political mobilisation and fragmented society that was in turn treated by elites as a series of competing communities and vote banks, further breaching the secular norm.

Chapters 3 and 4 document how Mrs Gandhi sought to restore Congress after 1967 by trying to move to the left, and then when blocked by interests situated within the state Congress machines, by successfully splitting the structures of the party. This was not a coherent strategy, but rather a series of contingencies undertaken by Mrs Gandhi and a cohort of advisors within the state as events unfolded. Most important of all, it was undertaken in the context of a threat to Mrs Gandhi's leadership by Congress chief ministers. Centralisation begged the question of how to directly mobilise the poor at the same time that the split destroyed the integrity of the Congress as a party, and exposed its over-reliance on local institutions of the state. How could Indira Gandhi use the radicalism of her politics to reconstitute the Congress as a mass social movement? And how in turn, could this reinvigorate the state at the local levels? This is a paradox that in almost every sense Mrs Gandhi failed to resolve. Congress appealed to the poorest and most numerous elements of society at a time when it was least able to deliver any of its electoral commitments to them. It lacked an ability to reconstitute the party to enable the poor to participate in existing forms of constitutional democratic governmentality.

Reliance on the state itself to step into this particular vacuum prefigured the shift towards populism, authoritarianism and policy failure. In retrospect, 1969 is as politically significant as the Emergency itself. It ushered in an extreme version of what is more generally known as the 'deinstitutionalisation' of the Congress system, as the prime minister pursued an attempt to create a truly centralised national party, but also deployed the state itself as a means to mobilise and direct the poor and control the party. It inverted the relationship between the prime minister and the chief ministers, compromised the Union presidency, and destroyed the integrity of Indian federalism. As Kaviraj noted 'it seems in retrospect that the systematic destruction of the party apparatus was not contingent; it was a necessary part of the populist transformation of Congress politics' (Kaviraj 1986: 1699; Kothari 1983b: 31).

Chapter 5 looks at the specifics of the Emergency through to the 1977 elections and the restoration of Mrs Gandhi. It seeks to offer a definitive account of this event from the perspective of the orthodox deinstitutionalisation debates, but also with specific reference to the state–society literature. From the original thesis the book also places in the public domain insights and interviews carried out in 1985 and 1986 with a whole series of principle actors, including S. S. Ray, Om Mehta and P. N. Haksar.

Chapter 5 identifies the consequences of the Emergency regime for the state in India, and identifies the societal impact the RSS–BJS had, both in the run up to the declaration, and in the aftermath of the 1977 elections. The collapse of the Emergency regime in 1977 is as extraordinary as its declaration of 1975. It marks a sudden U-turn in the collaborative strategies of Mrs Gandhi and a retreat into electoral politics as the only viable alternative to continuing the extractive and negotiated elements of state power. The attempts to reconfigure this process under the auspices of the 42nd Amendment – to make essential elements of the Emergency institutionally permanent – failed dramatically. Instead of cementing the Congress within the state and sanctioning further attempts at dominating society, the 1977 elections shattered the hegemony of the Congress and its electoral strength in parliament. The result was the Janata phase of 1977–1979, a phase that exposed the state to ongoing and dramatic democratising forces that had been working throughout society, galvanised both by the Emergency and their opposition to centralisation. It is in this radicalising and differentiating context, the democratic 'big bang' that followed the collapse of the Emergency, that *Hindutva* would come to political ascendancy.

Chapter 6 begins by reiterating the aspect of the societal reaction and resistance to the Emergency regime, refocusing on the JP Movement and the involvement within it of political and cultural forces associated with the BJS and the RSS. In this I have made use of new material provided by Christophe Jaffrelot (Jaffrelot 1996) and others working on Hindu Nationalism generally that have demonstrated the significance of such resistance to the later credibility of the Sangh as a whole. The Emergency provided an ideal opportunity for the RSS to consolidate its position within society, and to expand dramatically during the Janata phase, through new social and cultural fronts, and through the reactivation of the Vishva Hindu Parishad (VHP) in particular towards the construction of a Hindu votary.

This chapter reconsiders the role the Hindu nationalists played in the formation and direction of the Janata government, and the reaction of the Congress-I government, under both Indira Gandhi and her son, Rajiv, to the issues articulated by *Hindutva* by the late 1980s. In the aftermath of Mrs Gandhi's assassination, a new political landscape emerged in which secularism – not socialism – becomes the dominant motif for electoral contestation, and one in which the BJP emerges as a focus point for an attempt to construct a viable national strategy aimed at capturing state power. Paradoxically the BJP's emergence is disguised by the extent of the Rajiv Gandhi election victory of 1984–1985, in which Congress-I had already mobilised communal and cultural anxieties close to the hearts of the Hindu Right in order to retain access to state power. In what context had the Congress shifted to court an image of a Hindu majoritarian community, even to the extent that the RSS considered it a better vehicle for Hindu aspirations than the newly formed BJP?

The weakening of the central state's determination to assert secularism affected all political parties and changed middle class discourse and debate significantly. Playing with communalism, appeasing it, even seeking to resist it in the form of the implementation of the Mandal Commission, merely consolidated the rise of the BJP towards its first 13-day government, and then eventual power in 1998. Along with Shah Bano, the decision to implement the Mandal Commission provides one of the two 'signifiers' of a new state–society dynamics. The Mandal Commission helped consolidate the urban, forward-caste constituency that favoured the BJP, since the Shah Bano case drove the same constituency away from the Congress. However the Mandal Commission did not achieve V. P. Singh's objective of creating a coherent other backward caste (hereafter OBCs) alliance that would support his populist socialist leadership of the Janata Dal; it rather fragmented an internally differentiated group that, in the context of India's electoral system, momentarily cemented the BJP to power.

Yet just as the language of order during the Emergency stimulated resistance and differentiation, the strategic use of *Hindutva* brought with it forces of resistance, both within the Sangh Parivar itself, and from the growing political consciousness of large sections of OBCs. Collaborating with the BJP after 1992 and the unruly demolition of the Babri Masjid risked further political disorder, while attempts by the BJP leadership to sidestep plebeian sections of the Sangh Parivar, and stress socio-economic issues of governance, risked alienating it from the RSS networks. When in actual coalition, and situated within the central state, the tensions within the Sangh Parivar would become even more apparent, increasing the BJP's vulnerability to vote-pooling by secular parties, and above all, a change in the attitude of the Congress towards coalition government.

From 1993 to 1994, the societal project of the RSS and the short term electoral logic of the BJP that so neatly coalesced in the early 1990s would begin to diverge. The increasing political cooperation of the *bahujan* would be a significant factor here, as would the assertion of a secular platform by the Congress. While the appalling events of Gujarat in 2002–2003 inject a note of caution, the book concludes that the 'impossible assimilation' of the early 1990s was the curious, paradoxical product of state–society relations in transition, and above all of rapid and uneven political mobilisation. It is unlikely, free of a sustained crisis, that *Hindutva* can be sustained as a political project without being repeatedly broken and fragmented through the state–society dynamics of twenty-first century India (Desai 2005: 254–63).

In conclusion, the book will return to the nature and extent of the disjuncture between political mobilisation and governance in India. It will summarise the argument that a change from *garibi hatao* to *Hindutva*, if viewed in a broader historical perspective, and with an eye to the central dynamics of state–society relations, is less surprising that it at first appeared in the late 1980s and early 1990s. India watchers should not, in retrospect, have been as surprised as they were over the rise of the BJP and the rise of Hinduness. Hansen drew attention to the paradox of *Hindutva* by asking how it was that, with all its unpleasant and extremist connotations and ambiguities, it had become consolidated within the heart of a middle class otherwise associated with the virtues of social capital, the bastions of civil society and the success of the world's largest democracy. He went on to add, as a corollary of this, that 'one of the most remarkable features of the entire phenomenon of Hindu nationalism is the relative ease with which it has been fitted into most of the authorised discourse on India' (Hansen 1999: 7).

In seeking to bring forward my analysis of the Emergency to suggest ways in which its authoritarianism and statism led to the rise of *Hindutva*, the features noted by Hansen are probably less remarkable than we all at first thought. What I have suggested throughout this text is that, shorn of their particular differences, the rise of *Hindutva*, like the Emergency, is a crises of government provoked by elites seeking to manipulate and control rapid, indeed unprecedented, political mobilisation.

The book concludes by exploring further Chatterjee's innovative but problematic embrace of social movements as emancipatory, when in fact they can be viewed as just another form of corruption, extremism and intolerance. Like many recent post-structural and post-modernist commentators, such uncritical support is in part a consequence of an a priori condemning of the modern state project, and of seeing in the state the source of all contemporary societal ills. In that political elites may manipulate societal forces and distort them for short-term gains, this observation is true. Yet in many senses the state is required to actually resist ideologies of intolerance that emanate from society and suggest themselves as legimating devices of rule. It is in this regard that the politics of leadership are again stressed.

There is a further irony here. During the dark days of the Emergency, the real threat to Indian democracy seemed to emanate from the state, and much analysis went into identifying and predicting the nature of this threat and how to overcome it. As the societal forces associated with *Hindutva* and the RSS came to play a critical part in redefining state policies, it was interesting to note a switch in the debates over the likely origin and route of extremism (or even fascism) in India. It was now as likely to come not so much down from the ramparts of the state, but up through the messy, emergent and often contingent forms of social mobilisation and social movements that, as well as arguably being progressive, are as equally likely to emerge, in Gurharpal Singh's telling phrase, 'exuding xenophobic fundamentalisms' of their own (Singh 2003: 225). And this does not simply apply to middle class conceptions of order and status. Only if state–society relations are meaningfully conceptualised together can the threats to democratic government and society be adequately identified and contained. The state may not just be central to the 'idea of India', it might be a valuable and useful line of defence of democratic governance.

Bristol 2006

1 Emergencies, states and societies
The study of Indian politics

The Internal Emergency, declared in the early hours of 26 June 1975, was for many a turning point in the politics of India. Despite a growing sense of crisis throughout 1974–1975, and amid general discussions over the prospect of a coup or some form of extreme government action, the event itself was a complete surprise. Yet the ending of the Emergency in March 1977, and the subsequent electoral landslide for the Janata Party, were equally surprising, if not bewildering, to those who had come to support the Emergency regime and who believed that it had, under the auspices of the 42nd Amendment, become in effect a permanent feature of political life. The extraordinary victory of democracy over dictatorship, and the sheer decisiveness through which the Indian electorate reaffirmed its commitment to an elected parliament, gave the event widespread international coverage, and became part of the mystique of India as the world's largest democracy. Mrs Gandhi was unseated in the constituency of Rae Bareilly, and over 22 members of the council of ministers also lost their seats.

Looking back on these events through the prism of Mrs Gandhi's impressive comeback in 1980, her assassination in 1984 and the emergence of the Bharatiya Janata Party (BJP) in power, the Emergency has become displaced as a decisive event, and overlaid with differing concerns and anxieties; other emergencies as it were. These stem now from the sheer level of corruption and incompetence of government (especially under Rajiv Gandhi and later Narasimha Rao) to the onset of coalition government and executive instability, and to the growing praise or critique of *Hindutva*. Ironically many of the concerns over authoritarianism, even fascism, expressed during the 1970s that focused on the dangers stemming from the Indian state have been inverted during the late 1990s to concerns over fascism and social violence emanating from within civil society (specifically cultural and religious movements rather than political institutions). Indeed the lament is not so much the threat of the state to civil liberties but of the inability of the state to resist social movements aiming to subvert and undermine secularism. It seems difficult, even irrelevant in these circumstances, to reclaim the Emergency as a turning point in the contestation of political, economic, let alone cultural identities in India.

In 2003, Emma Tarlo noted that 'the Emergency occupies an unusual place in the Indian past. It has been much mythologised but little studied. Too recent to be of interest to historians and yet too distant to have attracted the attention of other social scientists, it has somehow slipped through the net of academic disciplines' (Tarlo 2003: 2). This observation is both accurate *and* incorrect. It is incorrect in that by 1998, over 1,200 academic sources existed that referred to the Emergency and sought to account for it either as an event in its own right, or to analyse its significance to Indian politics generally. It is accurate however in pointing out that many of these works, along with the events they sought to articulate, have become part of a temporal anomaly, dated in that they speak little to pressing events from the 1990s, or they have become part of a dated academic discourse that has simply not stood the test of time.[1]

This chapter aims to review some selective works on the Emergency, to argue that they have retained their relevance to a study of Indian politics and the rise of the BJP in the 1990s. I have also sought to bring many of their specific insights forward to debate secularism and culture, two aspects that were peculiarly absent from the earlier literature. It is my argument here that the Emergency retains its centrality because as an event it accounts for the dramatic decentralisation of national politics in the wake of the 1977 elections (despite the apparent Congress landslides of 1980 and 1984), and it provides the entry point for the BJP–RSS combine to mobilise societal support around *Hindutva* and to begin to construct a Hindu vote. In Hansen's terms, the Emergency marks the first visible moments of a disjuncture between ongoing political mobilisation and the institutions of government. It is within this disjuncture that the political, social and cultural terrain of India in the 1990s emerges as a site for the rise of *Hindutva* (Hansen 1999).

Not only has the Emergency been overtaken and obscured by subsequent events, but also by the earlier language and methodologies used by academics to analyse it (Toye 1976: 303–20). In retrospect, looking at the first summaries of the literature, there is a particularly structural, indeed often overtly deterministic approach to these studies of the Emergency and its wider significance to Indian politics (Mayer 1984: 128–47). Almost no study appertaining to the Emergency mentioned either culture, or indeed secularism, until the mid-1990s. Partly this was because the crisis itself occurred before Foucault's hegemony became established within social studies and at a time when many Indian scholars were still influenced by structural Marxism and forms of dependency thinking. Even to non-Marxists, the study of Indian politics was premised on a Crickian dictum that politics was bound up with the access to, exclusion from and the nature of, state institutions and the regulation of conflicts that were defined as overwhelming material (see Zavos *et al.* 2004: 1–15). Only by the mid- to late 1980s, and arguably even later, would there be a decisive displacement 'from politics to discourse, materiality towards culture and class towards community' (Bose and Jalal 1998: 8). Somewhat ironically, it was only with the rise of *Hindutva* itself that the definition of politics would become sufficiently broadened to deal with 'competition over symbolism and the strategic location of cultural signifiers – a competition of style and performance – even over what constitutes a public space – as much as over jobs' (Zavos *et al.* 2004: 3).

Bringing together these two events (the Emergency and its immediate causes, *Hindutva* and the rise of Hindu Nationalism) through a methodology that looks at state–society interaction and examines both through the prism of social mobilisation, requires a complex summary of two sets of quite different literature.

Initial studies of the Emergency were dominated by attempts to relate the event to developments within the Congress as a national party, the specifics of the prime minister, and then more indirectly the nature of the Indian state and social interests. Many of the more immediate commentaries did not focus on the state at all, and certainly did not offer to explicitly theorise the state as such. Before setting out in bold terms the dominant methodologies used to analyse the Emergency, and to highlight their contributions to the approaches I have adopted here, it is necessary to set out a crude, thumbnail sketch of the events of June 1975. The sequence of events is, for the moment, set aside, although the way in which the drama unfolded was not without obvious consequences for subsequent analysis.

In 1971, after a dramatic split with the Congress Party 'bosses' in 1969, Mrs Gandhi called a midterm national election – India's first – and went to the people of the country with a radical manifesto promising to abolish poverty and to deliver a socialist India. The *garibi hatao* (abolish poverty) campaign produced a large electoral majority for Mrs Gandhi, and was seen to confirm her earlier realignment of Congress to the Left in the company both of the Communist Party of India (CPI) and a group of radical young Congressmen, situated within the Congress Forum for Socialist Action. The crisis of 1969 – ostensibly over the election of the Union president – had been over ideology, but it had also been about Mrs Gandhi's continuation as prime minister.

Between 1971 and 1974, this massive majority, augmented by a series of state elections held in the wake of the Bangladesh war, was perceived by many to have been frittered away. The Congress Party – profoundly damaged by the split from the national to the local level – was forced to improvise the institutional structures necessary to implement agreed policy and to monitor and report back on the effectiveness of implementation. Such 'ad hocism' often placed power in the hands of social groups that had no interest in implementing *garibi hatao* policies, or had no skill to do so. Added to this disintegration of the machines of government and party was a growing economic crisis (involving the first international oil shocks of 1973 and two poor monsoons) and a series of corruption scandals involving prominent Congress chief ministers, first in Bihar and then in Gujarat. By 1974, political opponents to Mrs Gandhi were articulating widespread resentment against her government, first in the states, and increasingly at the centre. The emergence of the JP Movement – a Gandhian-inspired protest – aimed at removing corruption and undertaking a national renewal involved members of the Hindu National parties and former members of the old Congress guard, many of them Hindu traditionalists who had opposed Mrs Gandhi in 1969, whom she perceived to be fascists. The clash of interests was read not just as opposition to her government, but to Mrs Gandhi's explicit commitment to socialism and the poor.

Within this general context of crisis and drift, three specific events crystallised the declaration of the Internal Emergency. The first was a decision by the Allahabad High Court, which found Mrs Gandhi guilty of two charges of electoral malpractice during her 1971 campaign in Rae Bareilly. Fifty other charges were dismissed. These had been filed against Mrs Gandhi by the defeated socialist candidate, Raj Narain, and had been progressing slowly through the courts for nearly four years. The court ruled that she should be debarred from her constituency and from holding public office for a specific period of time. The government immediately appealed to the Supreme Court for an unconditional stay until an appeal could be heard in full. The opposition called for Mrs Gandhi's immediate resignation.

The second event was the electoral defeat of Mrs Gandhi's Congress Party in the state of Gujarat, the news of which arrived on the same day as the court verdict. The election, delayed under President's Rule and then conceded in the wake of a Gandhian fast by Mrs Gandhi's arch rival, Morarji Desai, was a major setback, and emboldened the opposition to ultimately target the national government in New Delhi. Given her domination of the campaign, a Congress defeat seemed to undermine her claims that she was indispensable to the party and to the nation. To many Congressmen, contextualised by the news of electoral defeat, the court verdict seemed to imply that she was in effect a liability.

The third and final event, emerging in the context of ongoing oppositional agitation to remove her from office through public protests and Gandhian-inspired non-cooperation, was the decision of the Supreme Court, made on 24 June, to issue a partial stay to the lower court's verdict, and retain some constraints on the prime minister's official duties.

It is at this stage that members of her own party, and indeed her own cabinet, began to express disquiet at the prime minister (Mrs Gandhi) continuing in office in a situation where she could neither vote in parliament nor draw her salary as a parliamentarian. On the evening of the 24 June, the opposition, organised in large part by the RSS and the BJS Party, held a mass rally and called upon the police and the army to disobey orders emanating from a corrupt and illegitimately constituted prime minister. It was this singular indiscretion by the opposition, intercepted and reported to the prime minister by the Intelligence Bureau, that Mrs Gandhi used as the official justification to declare a state of Emergency under Article 352(ii) of the Indian constitution. Beginning with a radio broadcast to the nation on 25 June, Mrs Gandhi sought to reiterate her socialist credentials by linking her actions to the necessity of the *garibi hatao* mandate, and by accusing the opposition of consisting largely of reactionary, right wing elements trying to frustrate her own radicalism and the progressive policies of the government.

Initial attempts to comprehend and analyse the shock of the declaration concentrated not only on these immediate events, but also the ways in which, both institutionally and constitutionally, the Emergency was possible and how indeed, since the 1960s, it had in some ways become more likely. Moving way from the official 'justifying' responses (Rao and Rao 1975) that by definition stuck closely and descriptively to the main events of 12–24 June 1975, more nuanced and academic analysis started out by examining the ways in which the prime ministership functioned under Indira Gandhi in ways that were increasingly incompatible with democratic institutional politics. Emerging quickly in the wake of the declaration, they emphasised the ways in which she had during her period in office actually altered the institutional framework of the party and government; or had simply destroyed it. To account for this fully, analysts need to go back to the mid-1960s when Indira Gandhi became prime minister. Above all, the Congress split in 1969 becomes crucial for understanding developments in the 1970s.

W. H. Morris-Jones's work is perhaps the best example of a systemic liberal approach used to analyse the Emergency. The motives behind Indira Gandhi's decision to react firmly to the agitations arose from the long-term changes she had herself brought about within Congress and the instruments of decision-making. The reasons cannot be found in the court case *per se* since Morris-Jones accepts that the charges were so minor as to be irrelevant under normal circumstances. However, the circumstances were highly charged and problematic. She could not stand down until fully cleared by the Supreme Court because here the risk was that rival elements within the Congress would squabble, or rather unite and prevent her from stepping back in again (Morris-Jones 1977: 20–41). An understanding of the reasons behind such growing factionalism within the Congress compels us to look at the actual changes 'brought about in the processes of government since she became prime minister and particularly since she emerged unchallenged after the splitting of the Congress Party in 1969' (Morris-Jones 1977: 27).

The style of Mrs Gandhi's leadership was affecting the institutional balance within the Indian political system through the extension of the prime minister's secretariat and the politicisation of the bureaucracy. The extension of the prime minister's power over that of the Union president and the AICC (All India Congress Committee) was to reduce the strength and independence of the chief ministers and their respective state-based parties, ultimately folding the 'government into the party'. Morris-Jones concludes that the net result of this process was the domination of Indian politics by a small, unrepresentative group, preoccupied with their own survival and oblivious to the more pressing social realities of India as dramatised by the demands of the JP Movement. By the mid-1970s these small groups had come to occupy critical positions within the government, to man several crucial central ministries, and to dominate the Youth Congress. Once this perspective is established the Emergency becomes less an aberration than a culmination of trends going on within the political process, a political elite and the institutions of government.

Manor in essence agrees with the the ideas of Morris-Jones and their differences are more of emphasis. Manor examines in some detail the ways in which the workings of the Congress electoral machines broke down, with prime ministerial nomination replacing democratic election. Focusing upon Indira Gandhi's continual intervention into state politics from 1969 onwards, Manor argues that she was increasingly going against the logic of the Indian system that evolved prior to the coming of the British. This logic was the necessity of running subcontinental politics within a decentralised system based upon policy compromises between the centre and the provinces, and between the institutions of the state and the co-optation of key societal elites as collaborators. Significantly, mass democracy changed the form, but not the logic, of this collaboration.

In his articles 'Indira and After' (Manor 1978: 318) and 'Party Decay and Political Crisis' (1980), Manor argues that, although the foundations of the Congress Party lay in the landed castes, its organisational strength and electoral success lay in its ability to form large, loosely articulated coalitions that incorporated regional and district derived patron–client relations that used local knowledge and often local idioms of power. Indira Gandhi ignored this logic when she went about centralising the party after 1969 and later collapsed the party and government together under her control. The significant split in the Congress of 1969, involving as it did some ideological elements, removed the old guard from the party and finally dispensed with the decaying roots of the Congress system left by Nehru. These activities by the prime minister had far-reaching consequences, from the adoption of central policy to the raising of party finances: 'the importance and style of fund raising since the early 1970s has brought about a change in the recruitment priorities for the Congress. The ability to deliver money to the Party coffers had taken precedence over persons linked with popular support' (Manor 1978: 785–803).

In failing to comprehend – or deliberately ignoring – the logic of a complicated diversified polity, Indira Gandhi's attempts at centralisation had exactly the opposite effects of those intended. The centre became increasingly isolated from the rest of India at the same time that it assumed greater formal control. Faced with increased central intervention into state affairs, Indian state politics increasingly differentiated along regional and cultural lines. Far from ensuring the political homogeneity of the subcontinent, Mrs Gandhi's centralisation actually furthered regionalism and the independence of the states through encouraging active non-compliance and resistance. Coupled with the language of socialism and empowerment, Mrs Gandhi encouraged and emboldened a broad social constituency who identified with being poor and with the promise of state aid, only to be frustrated and politicised by policy failure and growing societal competition.

Kochanek, like Manor, believes that an understanding of the Indian Emergency requires a wide political and historical perspective, in his case an understanding of the distribution and separation of constitutional power under the British. Here the emphasis is not so much upon diversity and parochialism but upon the unitary tendencies of the Raj. Indira Gandhi was able to re-recentralise the Indian federal system simply because the constitution allowed her to. Once Mrs Gandhi had managed to dominate the party, the Working Committee and the Parliamentary Board, she could not only control the nomination of chief ministers but, given the domination of the central legislatures by the Congress, the election of the Union president itself. Although titular, the folding of the Union president into the patronage of the prime minister gave Mrs Gandhi access to further executive authority beyond the immediate and indeed effective recall of parliament (see Hewitt 1994: 48–73). In this specific sense, the Emergency had been made possible by the consequences of the Congress split of 1969, which had placed the Union president – the head of state, and with all the formal powers that this entailed – within the ambit of the prime minister's patronage. Yet behind this lay a peculiarly colonial format of authoritarian democracy in which executive power was removed from the immediate logic of representation.

Notions concerning the decay or decline of the Congress institutions predominate in American (or American-inspired) writings on Indian politics in the late 1980s (Weiner 1982: 339–55, Bates and Basu 2005). Their strength lies in their ability to grasp India's degree of institutional sophistication, but none set out to overtly theorise the state at all. Neither Manor nor Morris-Jones makes explicit reference to the relationship between the Indian state and social classes, although both are keen to emphasise the inability of government to carry out various policies, and question how socially representative or how capable Mrs Gandhi's government was in carrying out and implementing policy on behalf of the poor. A majority of academics analysed the Emergency from the basis of those who stood to gain from it. In short, and central to the liberal view, Mrs Gandhi stood to gain the most in that the Emergency was a direct response to the court case aimed at removing her from power, either directly, or by so paralysing her that she would be removed by her own party. In broader terms, particular groups were certainly benefiting from the Emergency, and despite her talk of socialism and of using the Emergency to renew her socialist mandate, Indira Gandhi clearly found support for her actions amid elements of the Indian industrial classes and large business houses. However, there was also anxiety on their behalf as to how far the Emergency would take a socialist turn, and as much anxiety on behalf of the government as to how far such industrially based, capitalist interests might benefit from the absence of parliamentary democracy and labour rights. The poor, too, had their growing concerns by 1976: fear of arbitrary arrest and intimidation, the threat of forced sterilisation as well as the demolition of their homes, and a political system from which they were now totally excluded.

J. D. Sethi (1969, 1974) in his two works published before the Emergency examines the struggles between Mrs Gandhi and the Congress chief ministers that culminated with the splitting of the party in 1969. Central to his argument is how Indira Gandhi utilised emerging, previously subordinate, class and caste forces in her private struggles against the chief ministers without accommodating these forces within the Congress itself at the local level. She manipulated, as opposed to represented, formerly submissive social forces. By the late 1960s, the power of state-based leaders had been undermined by rapid socio-economic mobilisation. Such changes had weakened the earlier established patron – client networks of electoral politics and threatened the Congress with electoral defeat and India with political instability. When personally threatened by the chief ministers, Indira Gandhi directly appealed to these increasingly assertive sections of the electorate on the basis of radical electoral pledges, pledges subsequently discarded once her position was secured.

This instrumental approach to electoral politics is, for Sethi, the essence of Indian populism, a term that many analysts associate with Mrs Gandhi, and who often discuss this in comparison with Europe and Latin America. Without significant political restructuring within the Congress, and without real economic concessions to increasingly assertive subordinate classes, the consequences of such populism lead to a sustained political crisis in which the 'structure of power is most likely to become increasingly dysfunctional, unproductive, anarchic and violent. Above all, it will make the constitution unworkable' (Sethi 1969: 140). This culminates in the abandonment of formal democracy by the state as it tries to close down institutional access to the leadership, or reconfigure access through a form of parliamentary dictatorship.

Rajni Kothari also sees the Emergency as part of the long-term, systemic difficulties of the Congress in facing up to democratic government. Kothari stresses the inadequacies of a narrowly defined, out-of-touch elite, entrenched in power and no longer responsive to the newly awakened social interests throughout India that have been empowered and mobilised through parliamentary participation. For Kothari, the root of this unresponsiveness, although to an extent to do with the style of a particular prime minister, lies in the history of an Indian elite and the related aspects of an elitist education and a bureaucratically dominated government. Significantly, Kothari returned to this theme in the 1980s in seeking to analyse what he saw to be the final demise of the Congress, and the wider dangers that arose from the mobilisation of India's masses within an increasingly decaying and moribund political system.

In an article written in the wake of the collapse of the Emergency (Kothari 1977: 32–42), Kothari saw the future of India's democracy in much the same terms as Sethi's contradictions between an increasingly aware electorate and a moribund system of political representation. Later, he noted somewhat prophetically that 'if the process of [politicisation] reaches the lower castes without in fact providing the institutional channels for challenging the elite hegemony, caste and communal identities will begin to operate increasingly outside the institutional space and become unmediated, direct and violent' (Kothari 1983b: 31), paving the way for profound disaffection, a sense of regional separatism and ultimately political violence as the state seeks to co-opt them directly. Kothari in part greeted the rise of coalition governments in the 1990s as a sign that electoral awareness was finally redefining the nature of the Indian political system, but the decay of the party system and the rise of a communalised political movement, such as *Hindutva*, was also for him another symptom of political decay, a form of political pathology, caused by state instrumentalism and a callous, indifferent political elite.

Lloyd and Susanne Rudolph have advanced a more sociological and historical view of the Emergency based upon several theories of 'class collaboration' and a more explicit version of the state based on what they refer to as state corporatism. Their view of corporatism draws upon a historical sketch of the formation of the state in India but in a wider comparative context of the rise of the modern state more generally (see Weiss and Hobson 1995). In their article 'To The Brink and Back: Representative Institutions and the Indian State' (Rudolph and Rudolph 1978: 379–400), the Rudolph's argue that Indira Gandhi was capable of consolidating her position through extending the powers of the state through her control of the Congress Party. Initially this involved mobilising the electorate behind an explicit socialist commitment to social and economic reform. However, faced with serious political resistance at the state and district level and, after 1973, increasing difficulties of political accommodation, the prime minister turned increasingly to a notion of executive democracy. Executive democracy involved attempts by the prime minister to use the institutional powers at her disposal to intervene within society and to bring together various societal interests under the direct auspices of the state – echoing the earlier emphasis upon state corporatism. This mediation was seen emerging in the early 1970s, when Indira Gandhi began calling for a committed bureaucracy, a committed opposition and a committed judiciary to help her establish a socialist pattern of society. Collaboration between the state and social interests did not prevent the state representing some interests more consistently than others, according to the wider societal power of those interests or rather the greater necessity for the state to take such interests on board. Writing just before the declaration of the Emergency, and then again just after the Janata victory, the Rudolph's were clear that, however indecisive the outcome, there had occurred a major change in the nature of the Indian state: 'what had been a version of the liberal State, in which citizens permitted the government the right to hold office and use certain powers that were constrained and counterbalanced through account-ability to the electorate, was largely if not wholly dismantled during the Emergency' (Rudolph and Rudolph 1977: 849).

Even after its reconstitution by the Janata government, this version of a liberal state was weaker and more reactive than before. In the mid-1980s, the Rudolphs presented a broad-based historical analysis of Indian politics, in which the decline of state autonomy under the compulsions of a peasant-based 'bullock capitalist' mode of production explained the Emergency as well as the continuing political stalemate within the Congress and the non-Left political parties (Rudolph and Rudolph 1987). Like Kothari, they were ambiguous in their analysis of the onset of coalition government and the ongoing regionalisation of national politics. While it could be seen to represent democratisation, it also represented ongoing state weakness and a crisis within the workings of the original constitution.

Francine Frankel presents, probably the most ambitious analysis of Indian political economy and the causes that underlined the drift towards the Emergency. Her work deals with issues common to the liberal approaches, stressing, for example, the institutional peculiarities of the Congress, and deals with issues relevant to the more explicitly Marxist works, such as the relationship between class interest and economic policy. Her book, *India's Political Economy 1947–2004* (Frankel 2005), is an expli-cit critique of the logic behind Indian planning. In it Frankel explores the class nature of the Indian political system and how, given the social basis of Congress' electoral support, it has failed to confront what are for her the underlying structural causes that contributed to the political unrest of the early 1970s and ultimately the Emergency.

The political crisis that made the Emergency possible involved the col-laboration of interests within the Congress Party and the resulting empha-sis, after independence, upon compromise and consensus. This involved the fudging of 'radical reforms' or the deliberate neglect of policy implementa-tion by the elite that would lose out if democratic institutions proved capa-ble of significantly redistributing social and economic assets. As political pressures increased on the government – pressures personally encouraged by the prime minister – the collaboration of interests within the Congress broke down and democratic governance became unworkable.

The actual Emergency remains largely coincidental to the wider class alliance elucidated by Frankel's work. Frankel never reduces Indira Gandhi or the Congress to the agent of one class. Rather, like Sethi, Frankel argues that the prime minister used the patronage at her disposal to create a popu-list style of leadership with the genuine belief that such popular pressure could, under central guidance, be used to push for economic reforms at the state and district levels. Mrs Gandhi is prevented from implementing the radical policies of 1971 and 1972 by other, more tenacious, interests than the small group of advisors around her. Like Sethi and Kothari, Frankel makes it clear throughout her work that she is addressing the success and failure of government economic policies vis-à-vis social interests. Nowhere does she construct a structural or even historical notion of the Indian state as a framework for identifying the deadlock within Indian planning.

Writing within their own tradition of seeing the state as something theoretically distinct from the mere institutions of government, and as something much more than a neutral arena, Marxist writers sought to outline particular accounts of the Emergency and to work them into broader debates about Indian politics and the peculiar capacities of the Indian state, or to locate a study of the Emergency within ideas of a post-colonial state which has a specific predisposition towards authoritarianism.

In the *Manifesto of the Communist Party*, Marx and Engels outlined the classic notions of the state–society relationship: 'the executive of the modern state is but a committee for the management of the common affairs of the whole bourgeoisie' (Marx and Engels 1977: 111). This statement conceives the state as identifying and acting in the common interests of the bourgeoisie as a whole, overcoming the tendency of the bourgeoisie to limit their concerns to the pursuit of their separate, immediate interests. Thus the modern capitalist state provided the vital institutional and coercive framework for achieving the coherence of the bourgeoisie vis-à-vis other classes. The institutions of representative government constitute the main form of bourgeois rule. However, under the dynamics of bourgeois democracy, the bourgeoisie grant to the proletariat access to government through universal franchise, and the capacity to form their own political parties and to participate in elections. This contradiction lay at the heart of Marx's analysis of bourgeois democracy.

The second strand to Marx's take on the state emerges from French revolutionary politics in the middle of the nineteenth century. In *The Class Struggle in France* and *The Eighteenth Brumaire of Louis Bonaparte* (Marx 1977: 96–166), Marx saw French politics as being not so much an expression of the differences between classes as the results of conflicts among the different fractions of the bourgeoisie itself. Factional interests developed within the French bourgeoisie to the point at which the interests of the class as a whole could not be reconciled with the workings of a bourgeois democratic state, but required a state capable of acting independently of any one faction of the bourgeoisie under an authoritarian leadership. Unable by themselves to secure political peace and security for their financial and economic operations and 'faced with the continual threat of violent outbreak which offered absolutely no prospect of a final solution', the French bourgeoisie under Louis Bonaparte abdicated the right to rule for the right to make money (the guarantee of stability over the price of electoral freedom) handing over power to an individual with his own idiosyncrasies of ruling: 'thus it happened that the most simple minded man in France acquired the most multifarious significance. Just because he was nothing, he could signify everything save himself' (Marx 1977: 237).

The Eighteenth Brumaire explored the events of 1848–1852 in France, and analysed circumstances in which the bourgeoisie would prefer authoritarianism to the stresses and difficulties of parliamentary democracy. Unable and unwilling to create the 'pure conditions of their own class rule', and unable to oppose the interests of the peasantry and the proletariat 'without the concealment afforded by the Crown' the bourgeoisie preferred the authority and stability of Bonaparte and the army to the political instability of the Republic and of electoral politics (Marx 1977: 423).

Marx contributes several other explanations to the political victory of Louis Bonaparte. He suggests that it arose out of circumstances in which the proletariat were relatively over-organised in relationship to their actual numbers. Themselves weakened through factionalism, the bourgeoisie as such exaggerated the overall threat of radical socialism. Thus Marx suggests that the authoritarian state steps in to manage where no class, or fraction of a class, has the capacity to rule society itself. Central to this peculiar dynamics, in which classes did not adopt their 'pure form of rule', was the existence of a broad peasant-based economy in which, although increasingly commercialised, peasant farmers had still not become divorced from the means of production to form a rural proletariat, and still required substantial state subsidies to compete against other societal interests.

Marx's observations on nineteenth century France raised important questions concerning the exact link between the form of the capitalist state and its political commitment to parliamentary democracy. It questioned the historical tendency wherein the development of the capitalist mode of production relied upon the opening up of the political system through electoral reform, and the ability of the bourgeoisie to pursue their own economic interests through representative political institutions. Of the many strands of theory within Marx's *Eighteenth Brumaire*, these two insights were to be advanced most frequently by later theorists attempting to analyse the social and political peculiarities of the state, especially in a third world or post-colonial context in which class polarisation was complicated by the continuation of large peasant societies and a bourgeoisie that was itself often divided between a nationalist faction concerned with the development of the national economy, and an international or comprador faction that remained interested in international capital and often remained associated with the former imperial power situated in the metropolis.

Post-colonial political elites became the contemporary equivalent of Marx's factionalised French bourgeoisie, with the nationalists being progressive but easily frightened by a radicalised working class, and a comprador class interested in foreign collaboration and, when necessary, a restrictive political system. This authoritarian faction was also prone to associate itself with pre-capitalist, predominantly agrarian interests who, in many cases, still dominated peasant-based economies with their culturally conservative populations. More importantly, it was willing to leave the externally imposed and often overdeveloped framework of the colonial state largely intact, and collude with the large and elephantine bureaucracy at the expense of internal reform and parliamentary democracy, in the main because it provides the necessary framework for order. Instructively, the alliance – or coalition – is profoundly contradictory, coalescing in the face of political disorder and sustained by crisis.

Such insights, again stemming from Marx's own colourful description of the Second French Empire as representing a well-fed and unrepresentative bureaucracy and a state that dominated all sections and factions of civil society, informed general understandings of a third world or overdeveloped state, some of which were applied to a specific understanding of India and the Emergency (Alavi 1973). Since then, theorising the Emergency as distinctive tendency of a third world or overdeveloped state has been the subject of prolonged criticism, both in general terms and also in terms of the specifics of the Indian context. In part the problem is a familiar one of agency. Mrs Gandhi and the court case and all the incidentals, which in some way structured the crisis, are marginal to the analysis itself. Moreover, identifying complex class structures were clearly determined political preferences is notoriously difficult, especially with reference to agriculture. Colin Leys pointed out that the centrality of the state was now a feature of the advanced capitalist countries as well, and so could not provide the basis for establishing a generically different third world state (Leys 1989: 39–49). Leys argued further that the colonial powers had no need to create powerfully bureaucratic law-and-order state structures because the weakness of indigenous social formations actually preceded colonialism, or more usefully had devised ways of collaborating with already dominant and emergent social interests during the colonial period itself (Bayly 1999). Where powerful colonial states came into existence they actually drew upon pre-colonial heritages, both institutional and ideological. What Alavi called the 'overdeveloped' colonial state was in fact constructed after independence with interests that might well have preceded the onset of empire (Bardhan 1984: 37).

Yet the idea that the Emergency represented, in some definitive sense, a state–society impasse in which the state takes on the attributes of private property, has been carried forward by a number of writers who have sought to free themselves from the overt determinism of third world state structures, while retaining an emphasis on the specifics of post-colonialism. Such works have, since the 1980s, referred to class and class alliances, but also made use of post-structural methodologies, increasingly those that stress the role of ideology and the concept of governmentality and autonomy. Pranab Bardhan, in a series of important articles, argues that the Indian state seeks, above all, to manage various conflicts between a 'dominant coalition' of propertied interests and between increasingly assertive subordinate classes empowered through democratisation. The task of attempting to coordinate this disparate group in a democratic fashion creates a political crisis and undermines the basic legitimacy of the state. The result is the autonomy of the central state declines. Bardhan draws attention to the economic strains within the dominant coalition and strains acting upon the dominant coalition as democratic management becomes more complex. External pressures are brought to bear through the workings of electoral politics and the ever increasing self-consciousness of the subordinate classes and their demand for inclusion. Apart from his explicit references to Marx, and the importance given to looking at the state, Bardhan shares much in common with Kothari's work.

In his article 'Authoritarianism and Democracy' (Bardhan 1978: 529) and in his collection of essays entitled *The Political Economy of Development in India* (Bardhan 1984), Bardhan argues that Indian political economy has been characterised by growing societal demands compelling a narrowly defined, state elite to act out and implement socialist policies without compromising their own economic and political interests. At the time of independence, this state elite 'inherited enormous prestige and a sufficiently unified sense of ideological purpose to redirect and restructure the economy, and in the process exerted great pressure on the propertied classes' (Bardhan 1984: 38).

Beginning in the mid-1950s, socio-economic change brought about more intractable divisions within the dominant coalition (between the propertied classes, the small industrialists and the emerging agricultural 'kulak' class).[2] Meanwhile electoral politics brought about a gradual increase in the number of economic demands being placed upon the coalition. Eventually the state's ability to accommodate all these various demands within its own socialist ideology falters, and when the opportunity (or necessity) arose, the political elite sought to close down institutional access to the state from society as a whole and switch to broad-based oppression while still articulating the language of socialist emancipation. Later these same societal forces would coalesce around *Hindutva*.

In an important work entitled *The State and Poverty in India* (Kohli 1987), Atul Kohli has developed some of Bardhan's views concerning the diminishing autonomy of the Indian state in the period after independence. However Kohli stresses the need to avoid socially reductionist views of the state and conceive state interests (or the interests of state elites or leaders) as in some cases quite fundamentally opposed to those of the dominant coalition of social classes – that is, to avoid reading a political elite within the state as a mere analogue of class interests within civil society.

Moreover, Kohli has extended Bardhan's emphasis on the importance and integrity of the institutional structures of the state, and the corresponding difficulties that confront a state leadership seeking to channel popular participation in economic and rural reform. Kolhi's work has done well to flesh out some of the actual empirical details surrounding democratic mobilisation in India during the period under discussion and to ground a view of the Emergency in a broader, nuanced study of Indian politics. His later works, that focused on the crisis of governability in India (Kohli 1990), foreshadow Hansen's conception of a disjuncture between the state and political institutions amid growing expectations and political mobilisation and an inflexible, isolated elite. Setting out to analyse India's growing crisis of government in a comparative context, Kohli noted in 1990 that 'sooner or later all developing countries become difficult to govern, and over the past two decades India has been moving in that direction. This trend contrasts with the situation during the 1950s and 1960s, when India was widely regarded as one of the few stable democracies of the non-western world' (Kohli 1990: 3).

Reiterating his earlier analysis in the circumstances of Congress collapse, Kohli again emphasised the decline of a 'legitimate' and modern state into a largely 'reactive' state, especially in terms of contradictory appeals to assertive, and contradictory elements of the electorate.

The ability of the liberal arguments to address directly the political aspects of the Emergency (the misuse of the prime minister's powers, the use of governmental institutions in the wake of the court case verdict) are clearly among their major strengths. While none seeks to exclude from its arguments the relevance of socio-economic circumstances, all are keen to emphasise the relationship between the dynamics of the Emergency and the changes in the institutional fabric of the Congress throughout the 1960s. This is particularly true of such writers as Morris-Jones, Manor, Kochanek and Kothari. All these writers develop the idea that institutions constrain, direct and channel societal communications while at the same time being acted upon and constrained by the often-divisive socio-economic interests of an unequal society. Within these institutions, leaders occupy specific spaces and locations that generate their own immediate interests.

Yet within a democratic polity, political leaders are directly subject to popular pressures. In response to such pressures, political leaders undertake electoral commitments that they cannot fulfil and which further encourage popular expectations. For writers such as Kothari and Sethi, this process of rising expectations has continued apace with the institutional collapse of the Congress and the erosion of its representative functions. The resulting authoritarian populism was the defining quality of Indira Gandhi's India. It explains the deinstitutionalisation of Indian politics and the political collapse of the Congress as a democratically constituted organisation capable of mobilizing popular support and implementing policy. But it is not clear whether these processes are contingent or structural. Without Mrs Gandhi's election there would have been no court case, without the court case there would have been no internal Emergency, but there would have been an ongoing political and economic crisis within the Indian polity: but it would have taken on a radically different form.

The weakness of many of the liberal arguments is that they lack an ability to conceptualise the state as transcending mere institutions of government, and to devise ways in which the state is compromised and contained by systematic societal interests, defined in some meaningful way by access to and control over the means of production, as well as by intra-elite conflict itself. Identifying state–society relations as an area of contestation has often been the preserve of Marxist and neo-Marxist writers, but has in turn been confronted by its own methodological issues. The state has now become one of the most problematic notions within current political and sociological writings. Attempts to conceptualise a theory of the third world state have been recently curtailed by serious controversies over the exact details of how the development of capitalism within third world countries is radically different from that in the West (Mouzelis). Ziemann and Lanzendoffer remarked as long ago as 1977 that 'before understanding what constitutes the peripheral state, we must first seek an understanding of the Capitalist state in general' (Lanzendoffer and Ziemann 1977: 143–177). Yet Gregor McLennan makes a more pointed reference to two essential problems confronting Marxism in particular on the matter of the state: 'that the idea of all states being solely instruments of class oppression must be doubted, and secondly that the role of the "capitalist" state is not simply one of facilitating capitalist interest. These are contingent questions to be decided by historical analysis, not philosophical assumption' (McLennan 1981: 99).

If political leaderships inhabit a clearly defined area in which they are frequently capable of ignoring economically determined class interest, it is not clear how the concept of autonomy as a specific element of crisis can ever be useful. The difficulties become not so much the weakness of particular Marxist scholarship but a essential failure of the Marxist paradigm. Problems in materially defining class, and of giving class priority as a political factor, are all incredibly difficult in the case of India, where caste and jati, language and religion provide niches of identity and potentially contradictory sites of political action. It is not just that the state–society relationship is problematic, so in turn are the relationships between religion, culture and production and accumulation. Mapping these routes, and imputing them as causal or primary, is a dangerous and difficult job for any analyst and often leads them to asking the wrong questions (Hansen 2004: 19–36).

To return to works specifically concerned with the Internal Emergency, Bardhan draws upon various other writers, particularly Skocpol, who have re-emphasised that the state has autonomous interests as an actor in its own right, 'going beyond the demands and interests of social groups, to promote social change, manage economic crises, or develop innovative public policies' (Skocpol 1979: 1).

In her work entitled *States and Social Revolutions,* Theda Skocpol sought to reclaim the state on the grounds of a more Weberian tradition as a set of administrative and coercive institutions, and argued that 'the state properly conceived is no mere arena; it is rather a set of administrative, policing and military organisations headed, and more or less co-ordinated by, an executive authority. Any state, first and fundamentally extracts resources from society and deploys these to create and support coercive and administrative organisations' (Skocpol 1979: 3).

But as Midgal points, out there are difficulties with this conceptualisation in that it exaggerates the unity of the state, and the extent to which is a unified, homogenous actor (Migdal 1988, 2001), suggesting an alternative conceptualisation of the state in society, in which the state as a myriad of institutions is crystallised by the dominant and transient forces active in society. As Mann puts it, political power 'crystallise[s] in forms essentially generated by the outside society [but] in forms that are intrinsic to its own political process' (Mann 1993: 56): in other words, in forms determined by its own history and its own distinct institutional dynamics. Midgal, like Mann and Skocpol, refers to the role of legitimacy in assisting the tasks of political leaders to act upon social relations and to use what material they find there for their own political projects. Specific cultural and religious idioms that can legitimate state action may well emanate not from the state, but from dominant or influential sections of society that the state needs to manage, contain, or ultimately, overpower. The overall emphasis here is on an institutional and constitutional framework within which state leaders seek to formulate and act out policies through collaboration strategies, with political elites as mediators over an uneven social structure and within an often chaotic state. The nature and dynamics of political change in India – summed up under the heading of democratisation – means that these elites are themselves subject to intense pressure to protect their own interests and to satisfy or at least control the expectations of broad-based electoral mobilisations through the state as well, but at multiple levels of state–society interaction and where forms of political power may well be different.

An important contribution to the general literature on India, as well as to later debates about the rise of Hindu nationalism generally, comes from Corbridge and Harriss's *Reinventing India* (Corbridge and Harriss 2000). What is significant about this work is its close association with – and in parts critique of – the ideas of Bardhan, Kohli and the earlier (slightly less post-structural) works of Partha Chatterjee (see below). While not seeking to abandon or ignore later discursive studies of the BJP, or the deconstruction of narratives emanating as justifications for the types and styles of political performance by the wider Sangh Parivar, these two authors claim to be 'unashamedly old fashioned…although concerned with the politics of identity, [the study] is founded nonetheless on a concern of class formation and the dynamics of accumulation' (Corbridge and Harriss 2000: xx). Central to these processes is the idea of the state and the need to conceptualise the continuing importance and necessity of a modern state for carrying out emancipatory politics through an active engagement with society and development.

Corbridge and Harriss draw freely upon the works of Barrington Moore, Skopol and Rueschemeyer to identify the state, in the first instance, as an institutional organisation characterised by a priori autonomy. India's political elite is narrowly drawn from a predominantly westernised middle class, and is dominated by the Congress, although (aided by hindsight on behalf of the authors) the prevalent ideas of a modern, secular western state subsumes – but does not eliminate – the alternative or closely associated designs of Hindu nationalism within the wider developmental discourse of India's leaders. In their analysis of popular democracy, Corbridge and Harriss follow the orthodoxy of the deinstitutionalisation debates, looking at the failure of the Congress to actually reform Indian society as opposed to merely consolidate specific elites, above all a dominant coalition of propertied interests. The dominant-coalition idea draws closely on Bardhan's work, and on Chatterjee's debates about the collapse of the Janata Party in 1979–1980 and the rise of a commercial, pro-market farming lobby that competes with industry and commerce for state inputs. In Corbridge and Harriss, this struggle facilitates the context in which *Hindutva* can challenge state power.

The component parts of this coalition consist of an industrial and commercial bourgeoisie, a growing rich peasant class (fluid in outline and identified through a series of dominant and emergent castes varying from region to region) and what appears to be a specific group of interests identified with the bureaucracy. This bureaucratic elite forms part of an intermediate class, made up of petty bourgeoisie interests and a growing ruralised small town lobby that emerge through the 1960s onward, which benefit from state employment, and the ability to manipulate and distort the regulatory structure of the state itself. Defined in opposition to classes and caste groupings excluded from the privilege of property owning, this coalition operates through a development strategy that enhances its own (ultimately contradictory) interests while trying to mediate growing pressures from outside the coalition, empowered through populist democracy. In keeping with the insights from Kothari, political accommodation is sought by the crude mobilisation of the intermediate and backward castes through the 1980s and 1990s, and not through either comprehensive land reform or concerted state intervention in education or social reform. As such it is a short-term, ad hoc compromise, determined by election as the sole legitimator of political power and one confronted by frequent crisis.

As the interests of the dominant coalition diverge (especially between the farming lobby and the commercial and industrial middle classes), and as democracy erodes traditional and vertical lines of authority, the state declines as an organisation and becomes more prone to violence and more incapable of mediating divergent regionalised and sectional interests. The Congress as a premier national party, and an implicit part of the state, rapidly collapses and in order to shore up its electoral legitimacy, it turns to a crude form of Hindu majoritarianism. In seeking a compromise with societal interests mobilised and in part empowered by *Hindutva* (along with other parties, including V. P. Singh's Jan Morcha), by 1989–1991 Congress enters into an open bidding competition for a Hindu vote. Herein lies the surprising shift in the nature of India and the apparent, swift compromise of secularism within the state itself. Yet the assumption of state power (in the regions first, and then by 1998, at the centre itself) generates its own conflicts within the Hindu nationalists and, in the face of ongoing political mobilisations within the Other Backward Castes (OBCs), drives a wedge between the RSS and the BJP.

While sharing much that is common with some Marxist and strictly non-Marxist analysts, what is significant about *Reinventing India* is that, it sets out a compelling understanding of the wider context in which the rise of the BJP becomes possible. One factor concerns the virulence of the rise of the farming lobby in pressuring resources from the state. Corbridge and Harriss argue that the so-called new farmers movements helped create conditions of mobilisation and direct agitations that were replicated for the communal and caste agitations and upsurges brought about by V. P. Singh's surprising (and evidently unilateral) decision to extend caste reservations to OBCs by implementing the Mandal Commission. In turn, increasing social violence between forward and backward castes, coupled with the collapse of the Congress, turned many elements of the dominant coalition to supporting the BJP, especially the urban and commercial middle class concerned with the plebeian incursions taking place around them.

Such interests were not particularly attracted to *Hindutva*, but in a specific context of instability, they could live with it. Just as some analysts argued that the Emergency provided the opportunity for the elite to abandon the messy, dangerous and unclean politics of democracy for the politics of order, much the same can be said for the utility of *Hindutva*. Corbridge and Harriss summarise these processes succinctly as a series of 'elite revolts' in the context where deepening democratic mobilisation has exposed the state to increasing levels of social violence. The decline of the central state is vital to an understanding of Indian politics throughout the 1990s. This is either a cause of celebration or a cause of deep concern. It is also, less subjectively, *the* precondition for the emergence of the BJP and its championing, at least in the early to mid-1990s, of *Hindutva*:

> Over the next quarter century [from 1964] there came a whole procession of governments which together presided over the decline of the state-as-organisation and which allowed the growing challenge of Hindutva to the modern state idea.
>
> (Corbridge and Harriss 2000: 68)

It is to this idea of *Hindutva* that I shall now turn. Beginning in the late 1970s, and in line with the revisions to state theory touched on above, the concepts of civil society and social formations – including class and caste – have also undergone a veritable intellectual revolution, by both Marxist and liberal alike. In part the causes were the same, a reaction to excessive determinism, but also a response to post-structural (and later still, post-modern) insights to the fluidity and subjectivity of social identities and new social movements. In the particular case of India, one of the most important inputs has come from a healthy and robust revisionism in South Asian historiography – exemplified by the subaltern studies approach and the critique of 'elite' Nehruvian nationalism. With this has come a revision of how the state is conceived and understood, but more profoundly, how social interests and identities are constituted and represented as part of the political process.

This revisionism has had the beneficial effect of problematising the concept of civil society to become, not a glib assertion of democratic capital, or the necessary precondition for a democratic state, but an area or space in which various interests – both material and cultural – mobilise and renegotiate the meaning of politics and the state itself. It is with these forces, and through a variety of means, that the state is compelled to negotiate in order to retain control and extract resources, creating an ongoing tension through which forms of governing and control are constantly invented and compromised. If revisions to state theory sought to free up the state from any a prior determinism with society, recent emphasis on social formations free them from being seen as somehow predetermined, dependent upon or even interested in state structures. Van der Veer has put the case more bluntly: 'I submit that we should take religious discourse and practice as constitutive of changing social identities, rather than treating them as ideological smoke screens that hide the real clash of material interests and social classes' (van der Veer 1994: xi).

It was within this emergent, intellectual landscape that some of the more fruitful studies and investigations of *Hindutva* came to be located – with its own agendas for social transformation. While I agree with this as a point of reorientating debates away from class determinism, I reject the discursive analysis that van der Veer uses, which actually replaced one set of essentialised identities with another. When van der Veer noted, in 1994, that 'current events in India reveal the continuing importance of religious nationalism' (van der Veer 1994: ix), his use of the term 'continuing' was moot to say the least. A more interesting observation might well have been to analyse why religious nationalism had suddenly become so important and central to national politics in India.

Until the mid-1980s, a majority of the writings on *Hindutva* and Hindu nationalism were either historical or confined to comparative sociology and the study of socio-religious reforms generally from the nineteenth century onwards. Political studies of the RSS and the role played by political parties such as the BJS were confined to texts such as Bruce Graham's *Hindu Nationalism and Indian Politics* (Graham 1990), or to regional or state-based studies of party systems, in which the BJS had been relatively successful. Only with the rise of Congress's communal posturing in the mid-1980s, the BJP's electoral breakthrough in 1989, and with the palpable sense of shock following the demolition of Ayodhya in 1992, did mainstream political analysis seem to take notice of the various components of the Sangh Parivar, especially the RSS, the VHP and the onward march of the BJP, and evaluate this as both a political and societal project.

As post-structural and post-modernist accounts of India weighed in, there emerged a growing alternative understanding of politics that stressed less the formal institutions of party and government than the wider informal dynamics and meanings attached to religious and cultural symbolism during the mass campaigns of the BJP and in particular during the run-up to the demolition of Ayodhya. Trying to understand how and in what ways idioms of 'weakness', or loss, became transmitted to the Hindus through sudden and mass religious conversions and the political machinations of 'the Muslim other' became an important part of understanding the dramatic changes that overtook India in the 1990s. Part of this revision involved a re-examination of what secularism in India meant below the elite level, and how it was carried out as a form of governmentality. This concerned not just the familiar post-modernist deconstruction of the modernist agenda, with its hidden assumptions about the power of rationality, the backwardness of religious identity and the inevitability of social progress, it also involved a series of more nuanced studies into how Nehruvian socialism – the idea of India that Nehru sought to articulate within the modernising elite – was shot through with contradictions and ambiguities. Here again, the contribution of subaltern studies was critical in identifying the role that religious and cultural symbolism played in the national movement and how such symbolism needed to be conceived anew, away from a 'sole emphasis on the bourgeoisie project'. These insights have been extremely revealing, even if they have ended up overemphasising religion, and ignoring the wider political economy in which religion thrives and reconstitutes the meaning of the sacred and profane in a dialectic with the state (Harriss-White 2003).

Shifting the focus away from the state, wider debates on the RSS and its social action – exemplified through its history of seeking to strengthen and transform Indian society – stressed its success in consolidating a change in the way that India's minorities were now viewed and how this had *influenced* the state itself through elections. Corbridge and Harriss, who identified this process as a series of elite rebellions in part over anxieties over social turmoil that continued in the wake of the Emergency, also recognised the more complex processes wherein elites sought to reintegrate economic gains with ascriptive, religious values by participating in and also redefining religious ceremonies and practices, the nature and scope of social voluntary activity and concepts of charity (van der Veer 1994). Given the weakness of the state discussed in the previous section, and its retreat from autonomy to reaction, the state became captive of these changes, even dependent on them, even as it sought to manipulate the outcome. This manipulation contributes – through its crude interventions and a narrow, electoral definition of legitimacy, to the further pathologies of the Indian state itself, and even to the redefinition and reconstitution of religion. In this regard, *Hindutva* is a thoroughly modern and elitist project, it is arguably not even religious, and its contestation might well empower not just a form of secularism, but a more pluralistic, tolerant view of Hinduism.

Partha Chatterjee's collection of lectures and talks, entitled *The Politics of the Governed* (Chatterjee 2004), acts as a convenient meeting point to examine some of these changes in our understanding of societal or social forces, not just in regard to where the state sits and acts, but with regard to the rise of Hindu nationalism and the cultural project of *Hindutva*. Chatterjee's own work on India has mirrored the wider shift within Marxist writings on state–society relations (sharing much with the language and definitions of Bardhan) towards a more open-ended, post-structural critical take on social and political identities, and their wider associations with modernity – best exemplified by his damning critique of the Nehruvian version of the secular state.

Indian democracy from the late 1940s until the crisis of the late 1960s was about the state *controlling* and moderating democratic politics, but as noted by many of the authors above, for Chatterjee the period from 1967 to 1977 marks a breakdown in the ability of the state to manage democratic governance through universal abstract ideas of socialism and state-led social emancipation. This 'crisis of governmentality' spawns new forms of resistance and representation that carry the prospect of genuine liberty, even if carried out in a form that is not immediately recognised as institutional and constitutional politics, or indeed even if it is pathologised by the state itself. Chatterjee places the Emergency itself as a 'key moment' in this ongoing crisis, when mass political mobilisation turned on the political elites who had sought to manipulate it, and brought popular demands into the 'proverbial corridors' of power for the first time (Chatterjee 2004: 49).

To fully understand what follows from these 'societal incursion' into the state (or more accurately into the politics of governmentality), Chatterjee makes a distinction between civil and political society. Civil society consists of the discursive and procedural space in which India's relatively small, metropolitan middle class act as citizens. Political society consists of the broad majority of Indians who do not participate within politics through the conventions laid down by the constitution, but through a whole series of paralegal, often illicit and highly flexible arrangements brokered by a series of collaborative elites. The elites require the legitimacy that comes through electoral mobilisation and support, but not necessarily the social and political costs of accommodating the demands of subaltern classes situated outside civil society. Accommodation can take place, but it is likely to be resisted on the grounds that it undermines the purity and etiquette of middle class politics, and risks more pollution and noise (Kaviraj 1997).

The belief that such accommodation will take place through the trickle-down effect and the educational and cultural inclusion of more and more of the subaltern classes into civil society is, of course, a central premise of modernisation, and indeed a justifying myth of the necessity of a modern, centralising, managerial state. It is just as likely, however, that, in response to mass mobilisation, civil society withdraws into itself and effectively seeks to manage the demands of the governed simply through the arithmetic of electoral politics and the minimum provision of social services and basic needs. It might also lead to the increase in voluntary organisations and the private provision through religious and cultural foundations of public goods – a powerful account of what has happened with reference to public policy in India during the 1990s. Chatterjee uses the provocative image of a reversion to 'indirect rule' which 'involves a suspension of the modernisation project, walling in the protected zones of bourgeois civil society and dispensing the governmental functions of law and order and welfare through "the natural leaders" of the governed populations. The strategy, in other words, seeks to preserve the civic virtues of bourgeois life from the potential excesses of electoral democracy' (Chatterjee 2004: 50).

The strategy is remarkably similar – if not the same – as *Hindutva*. Again in a slightly different language, we have reached Hansen's major disjuncture. Although grounded in a very different intellectual language, the imagery that informs much of Chatterjee's recent investigations into the potential opportunities – and dangers – of democratising political society in the context of the modern state are similar to many of the authors discussed about with reference to the state, the role and interest of a dominant coalition, and the tensions inherent within mass-based democratic systems like India since the 1970s. Yet instead of being tied to a specific moment of crisis (such as Mrs Gandhi's declaration of the Emergency), for Chatterjee these tensions are the product of the condition of modernity itself, the specific origins of the state in the post-colonial world and the heterogeneity (or pluralism) of post-colonial societies. Indeed, arguably, in keeping with other writers like Jalal, these crises are a product of the state itself (Jalal 1995).

Yet while Chatterjee is pessimistic about the likely accommodation of political society within an inherently exclusionary format of the state, he appears optimistic about the possibility that political society will, in being exposed to the governmentality of the state and attempts to diversify control, open itself to a process of internal transformation, which might well transform the nature of governmentality as well. There is more than a trace of the old Marxist dialectic at work within Chatterjee's work here, despite his critique of modernity itself, that enables him to look at innovative forms of political practice at subaltern levels ignored by more orthodox political analysts, even when they appear to appeal to violence and entail intolerance – a problem Chatterjee acknowledges but has yet to address (Chatterjee 2004: 76).

The re-conceptualisation of society–state relations, through Chatterjee's critique of Putman's debates on social capital and the largely atheoretical debates within the democratisation literature on civil society, becomes useful for locating the site on which informal and novel forms of political activity – linked to identity and mobilisation – take place (Bekerlegge 2004). These are the forms of symbolic, theatrical, discursive acts discussed by Hansen (Hansen 1999) and by van der Veer, but they do not take place in a vacuum – again the ubiquity of the state is underplayed. In his essay entitled *Populations and Political Society*, Chatterjee states 'civil society...represents, in countries like India, the high ground of modernity. So too does the constitutional model of the state. But in actual practice, governmental agencies must descend from the high ground to the terrain of political society in order to renew their legitimacy as providers of well-being and then to *confront whatever is the current configuration of politically mobilised demands*' (Chatterjee 2004: 41, emphasis added).

This idea of the state – as almost passively modelling itself on current social demands and subaltern styles of politics – constitutes a rather extreme idea of 'societal autonomy' from the 'state,' a novel approach, in that, we see issues and agendas emerging outside 'the proverbial corridors of power' compelling the state to adopt and change the language and the idiom of politics in order to manipulate and use them for its own interests and of specific elites within the state. Chatterjee has yet to explicitly set out an understanding of *Hindutva* within this theoretical construct of society – state relations, but part of his analysis leads him to recognise that it did not originate within the state *per se*, but through the waves of mobilisation unleashed through democratic mobilisation, and the defeat of the Emergency in particular. As such it came by the late 1980s to constitute the dominant *current configuration of politically mobilised demands*, and compelled the state to seek some accommodation with it. Yet is it *Hindutva*, or the state's attempt to manipulate *Hindutva* for means of electoral legitimacy, that constitute the current political crisis in India? To what extent did state forms of governmentality shape the saffron wave that later pressured it into a series of concessions? The tendency is for Chatterjee, and critics of the Nehruvian state and of secularism in particular, to blame the state.

It is not my intention here to review and clarify the many contradictions within state theory, but rather to combine many of the ideas discussed above into a useful, if not eclectic framework for analysing the Emergency and evaluating its place in the current study of Indian politics, dominated by the dynamics of democratisation and the politics of *Hindutva*. In seeking to gather together the various debates on state–society relations, I want to return to the works of Dietrich Rueschemeyer and his co-authors (Rueschemeyer *et al.* 1992), Theda Skocpol, and the broader historical comparative work of Linda Weiss and John M. Hobson, combined with insights from Migdal.

Such works propose a series of analyses of the continuous interactions between state institutions, civil society and social classes (pace the issues raised by Chatterjee regarding political society). These interactions are identified broadly and pragmatically in relationship to economic production, ownership and access to resources.

What Weiss and Hobson call the 'neo-statist' (Weiss and Hobson 1995) view is something that comes out of many of the authors cited above (Kothari, Bardhan, the Rudolphs and Kohli), even though they have not explicitly used it or cited the relevant specific literature. Its perception of an institutionally grounded, ideologically legitimated state is fundamentally anti-deterministic, seeking to locate the state within broader structures of social power on a contingent, particular basis. Furthermore, following Mann, and working within a well-established Weberian framework, Weiss and Hobson reiterated a version of 'embedded autonomy' in which the state cultivates collaborative strategies with civil society and the uneven and unequal distributions of power that exist within it, to mobilise and direct it towards given ends. In identifying specific aspects of state–society interaction under the headings of penetrative, extractive and negotiated, Weiss and Hobson outline a spectrum of activity that, in the case of India, can be empirically demonstrated with reference to the Independence period in particular, but also more specifically for the Emergency and for what followed in the wake of its collapse.

While these approaches are useful (and I think, capable of meaningful synthesis), they have tended to ignore the role and influence of ideological, cultural and religious factors that have been so usefully explored by the post-structural approaches. There is a need, where possible, to retain the insights offered here by Rueschemeyer's understanding of state–society interaction, while correcting the more materialist, reductionist bias within the study of politics (and democratisation) it has tended to offer. Yet their approach seems perfectly capable of accommodating, alongside materialist notions of extractive and negotiated power, the broader cultural and religious forms and identities that the state encounters and encourages through the act of governance, and through which power and the state are perceived, accommodated and challenged. Political elites fight through party politics to control state power. This elite is not a direct analogue of class or caste power, it seeks to manage these societal forces and the power they control through policy or rhetoric, and as so usefully revealed by post-structural scholarship, idioms and metaphors that enhance new forms of community and identity within the political process. So, too, do actual political leaders with their particular susceptibilities, their charisma and their private agendas.

It has been my intention in this chapter to review and draw together many of the studies on the Emergency, and to link them to current studies on *Hindutva* and the rise of the BJP from the 1980s onwards. This task has been made difficult by the sheer range of material, and by the diverse methodologies used. However, what it has revealed is that studies on Indian political mobilisation had converged around a shared concept of state–society relations managed through the ongoing dynamics of democracy, in which, a small and narrowly defined political elite sought to direct and manage a deepening political process. One of the managerial strategies used for this was a socialist one, seeking to bring about significant political, socio-economic and cultural reform. This strategy was to some extent embedded within the constitution and the logic of anti-British nationalism. However, the institutions of state and the then dominant Congress Party (fused together in a particular way) proved incapable of restructuring the state to significantly carry out the extractive and redistributive policies such a strategy required. This is the beginning of the critical disjuncture that prefigured the rise of Mrs Gandhi and the various intra-elite squabbles within the Congress up until 1977. With the collapse of the Janata Party, political elites fighting for state control confronted – in the useful phrase of Chatterjee's – *Hindutva* as one of the 'current configurations of politically mobilised demands'. Indira Gandhi was particularly prone to exaggerate the RSS, as indeed her father had done. The result was the switch to overtly pro-Hindu moves to manage it, further politicising it as a result. The move was then followed by almost all the non-communist parties. The result was the breakdown in what Jaffrelot has identified as the hegemonic use of secularism to prevent the open use of religious appeals within the formal institutions of the central state. Congress, under Rajiv, ended up in an open bidding war with a rejuvenated BJP for an ethno-religious Hindu votary, and more specifically, the VHP. Through shocks such as apparent mass conversions, the Punjab and Assam crisis, and the widespread mobilisation of Muslims in the wake of the Shah Bano case, Jaffrelot notes that the conditions came into existence, not witnessed since the 1920s, in which Hindus felt under threat, and in which societal forces could momentarily fall in behind a political party.

The argument here draws on this, and the insight offered by Corbridge and Harriss, that the decline of the state allowed the growing challenge of *Hindutva* to take hold. Only a weak state, and a leadership within desperate for power, would have shifted to a selective engagement with the apparent power and influence of religious – cultural nationalism, a new strategy that abandoned *garibi hatao* and sought to use and manipulate dominant social forces that were themselves reacting to the unstructured and increasingly violent waves of democratic mobilisation. Again, such a strategy was possible not just because of the apparent societal power of the Hindu nationalism, but because caste and class mobilisation of the poor was regionalised and fragmented by both the party system put in place in the wake of Congress collapse, and a first-past-the-post electoral system.

The 1990s, both in the run-up to the demolition of Ayodhya and afterwards, Indian politics is characterised by the strategic abdication of the state in seeking to set a clear agenda for social and economic policy. This abdication is defined through the language of controlling and restoring order as opposed to redesigning outmoded forms of governance. Yet once in power, the BJP and the VHP confront their own difficulties of consolidating state power amid the continuing rise of the OBCs and other subaltern classes, and amid middle class fear that the BJP cannot control its own 'god men' (Adeney and Saez 2005; Harriss-White 2003).

Hindutva is resisted as much as it is advanced by political mobilisation but in a context in which it cannot restructure or enhance state capacity to deliver policies – or to innovate on governmentality simply through the act of voting or the theatre of politics – the democratic disjuncture continues. In this regard, and in the final analysis, there is an interesting and curious parallel between the Emergency and *Hindutva*. Both are manifestations of the failure of political elite to respond adequately to political mobiliation. Both also expose – or draw attention to – the specific weaknesses of the origins of the state in India. Such weaknesses are not that of the state *per se*, but of its current configuration within society, and the weakness of political leadership.

2 State–society relations in India 1947–1950

A democratic polity?

To account for the particularity of the Indian state, and the societal context in which the state operates, it is necessary to understand three aspects of the Congress. First, the ideological aspect of a movement for national independence; second, as a political party mobilising voters; and third, as an organisation aspect of the state itself (see Brown 1985; Low 1977; Tomlinson 1993). Many commentaries have stressed the continuities between the colonial and the post-colonial period, with reference to British institutional reform both directed and in part driven by India's complex and highly pluralistic social structures. The emergence of Western-style democratic government also involved the co-optation and adaptation of the languages of liberalism and forms of representation, as well as their contestation by more traditionalist and radical elements of society. The Nehruvian conception of socialism and secularism – often referred to as *Indian* nationalism – conforms very much to a modern liberal strategy of government, while what has been identified as Hindu nationalism occupies one that has been described as authoritarian, homogenised, corporatist, even fascist. Both were to find a place within the Congress, with Hindu nationalism mediated through Hindu traditionalists and Gandhians working alongside Nehru.

This chapter will look at the emergence of these ideas of India, and how their ideological agendas became incorporated into the Indian state at the time of independence. It will look at the formation of the constitution in some detail, looking at how the document set up specific institutions drawn from the colonial experience, through which ideological projects could be implemented, especially with reference to secularism, federalism, social reform and economic development. The chapter will conclude with an examination of the Emergency clauses that were retained from the 1935 Government of India Act.

As differing strategies and ideological projects, Hindu and Indian nationalism emerged at around the same time between the late 1880s and the early 1930s. Somewhat paradoxically, both were thoroughly modern: and more importantly, difficult to isolate, despite the apparent dominance of the Congress from the 1930s onwards, and of Nehru within the Congress itself. Until the recent emergence of the Bharatiya Janata Party (BJP) and the crisis of secularism in India, there has been a tendency to marginalise the Hindu nationalists and their impact, via Hindu traditionalists working within Congress, on Indian politics and on the founding myths and institutions of the state. Corbridge and Harriss note that 'the Hindu Nationalists played an important part in the struggle for freedom from colonial rule, but their place and their influence was denied during the ascendancy of the modernising nationalist elite' (Corbridge and Harriss 2000: xviii).

Yet, as will be argued below, even the idea of a modernising, nationalist elite separate from, and indifferent to, Hindu nationalist concerns, is misleading. Since the late 1980s, academics and social activists have sought to demonstrate how, away from high politics and the bastions of urban-based, westernised India, the tactics and strategies of the Congress merged with and cohabitated with communal or xenophobic agendas. Until 1937, members of the Hindu Mahasabha worked closely with the Congress, and K. M. Munshi, Union Minister of Supplies in the first Congress government, worked closely with Hindu nationalists, and more closely with Sardar Patel, India's Home Minister. As a powerful backer of Hindu traditionalists, and associated with the radical outspokenness of Tilak, Patel was seen as Nehru's ideological rival within the Congress, and a challenge to a Nehruvian project at odds with many Hindu politicians. In 1948, while acknowledging clear differences between them, Sardar Patel had called upon the RSS and the Hindu Mahasabha to join with the Congress, evidently recognising their shared commonality.

Political theorists have often characterised Congress and Indian nationalism as a form of 'step-in-your shoes' bourgeois-based liberal movement, entailing the wholesale adoption of the superiority of western forms of governance. Christophe Jaffrelot, following Geertz's work on the ideological use of culture in colonial societies, has argued that the emergence of Hindu nationalism was a response to, and in part a selective incorporation of, western forms of organisation and ideas (Jaffrelot 1996: 11–75). This invention of tradition, of *Hindutva*, involved the formation of an ideological identity aimed at defending values and patterns of belief otherwise threatened by modernity and the representative state. The adoption of a territorial, multicultural nationalism by the Congress, and the commitment to a secular state, would become the principle battleground for an alternative version of society (and by implication, its particular relationship to the state) based on an ethno-religious Hindu identity that recognised Hindus as the majoritarian community.

However, there is a tendency to overdraw the distinction between the two ideas of India and to ignore the complex interconnections between their aims and aspirations. Through a convenient form of historicism, 'it is usual to prune Nehru's complex, changing views on religion into the now withered bush of "secularism". But this imparts a misleading ideological fixity to it' (Khilnani 1997: 176). Again, Jaffrelot has argued that Indian nationalism was able to contain the mobilising strategies needed to empower Hindu nationalism as a political project because, for a while at least, 'secularism had emerged as the legitimating norm' of the Indian political system, but this too exaggerates both its permanence and the level of ideological contestation between it and Hindu nationalism.

In part this contestation took place within the same elite, an expansive middle class that had emerged within the context of the British colonial state, and the construction of institutions of government and a discourse of a public space that defined political behaviour and action. V. D. Savarkar's elaboration of *Hindutva*, in 1923, which was to inspire the rise of the RSS, turned Hinduism into a form of Brahmanical culture drawing inspiration from specific Western authors and thoughts on organisation. M. K. Gandhi foreground religion in the first non-cooperation movement, seeing in his own plural ecumenicalism, a commitment to truth and freedom. Even Nehru, in constructing his own version of a multicultural India, could not entirely dispense with a history of Hinduism as a great tradition – even if ultimately he saw it was destined to wither in the face of modern, scientific rationalism (Kaviraj 1997). Although eschewing direct involvement with the state, and seeing to organise through society in order to bring into being the new Hindu, many people would join the RSS to be part of the national movement to liberate India from the British through a shared language of governance. John Zavos succinctly notes that such shared resistance and agitation 'illustrates the way in which the ideologies of Indian nationalism and Hindu nationalism still merged [during the 1920s]. They emanated from the same middle class base and attracted the same middle class people as activists' (Zavos 2000: 193). Although they would diverge under the ascendancy of Nehru, they would converge as strategies of governance during the extensive political mobilisations that took place from the mid-1970s onwards, by an embattled political elite that would use religious symbolism to sore up its legitimacy, but also arguably to reclaim its religious identity as part of, even in defence of, their new material wealth.

The Congress started as a Western-educated, urban elite passing off as an English political society linked to reform within the empire and overt expressions of loyalty. Between 1885 and 1920 it broadened its socio-economic base and the scope and nature of its political demands in response to widespread changes within the British Raj, the wider socio-economic dimensions of empire, and the transformation of Indian society. It decisively shed its elite mode of lobbying and petitioning for reform under the influence of M. K. Gandhi and the onset of the mass disobedience campaigns of the 1920s, although the use of mass appeals to dominant peasant castes and Gandhi's patronage of untouchability caused anxiety and concerns within middle class circles. Subaltern scholarship has repeatedly pointed out the control by Congress over independent peasant movements, seeking to bring them within the hegemony of ideas of representation and reform (Guha 1998). Although Congress adopted various forms of social organisation, its emphasis on elections and access to the institutions of the colonial state stressed moderation, not transformation through a social movement, a strategy encouraged by the British Raj. As such 'images of colonial organisation, exemplified by the state, became a bulwark of modernity' for socialist elements within the Congress, primarily Nehru himself, who saw the state as the principle means to implement socialism and secure freedom. Ironically perhaps, such images of colonial organisation and authority would influence the societal structures of *Hindutva* through the RSS (Kanungo 2002).

By the 1930s, Congress had gained the support of various dominant castes throughout the countryside, landowners from the traditionally higher castes (such as Brahmins, Kshatriyas and Banias), and where they had emerged, some dominant cultivating castes. The consolidation of Congress from the 1920s onwards came to rest on the co-optation of powerful castes and jati clusters controlling land and resources through representative institutions bequeathed by the colonial power, a strategy that immediately constrained the radical socialist elements within the party, and which would generate tensions in the immediate wake of independence when the commitment to a socialist pattern of society called for changes in the nature of the state and perhaps constraints upon parliamentary democracy. The desire of Congress for continuous dialogue with the British over reform involved the calling off of its own agitations when it appeared to jeopardise constitutional reform or initiate social and especially agrarian rebellion that challenged both its own elites and its image of a responsible and capable 'government-in-waiting' (Pandey 2006a: 119).

The use of counter-hegemonic strategies using symbolic resistance that seemed to be drawn upon traditional identities (best exemplified by the use of satyagraha and swadeshi), disturbed many westernised Congressites and were always carefully tailored to gain access to, and not overthrow, the state. Khilnani has argued that 'Nehru's acceptance of modernity as integral to the definition of a free India implied the need to turn the grand, complex national imagination into a state form' (Khilnani 1997: 171). Given that Congress' strategy of aggregation through the use of pre-existing notables (Corbridge and Harriss 2000) over direct and continuous social mobilisation, the party lacked an ability to reform society through party work (especially following the eclipse of the Gandhian critique of the state after independence) and thus, from the onset, relied upon the state as the principle instrument of change. While sharing an emphasis on organisation seen within the various Sangathan movements of Hindu nationalism, Congress's party structures would remain subservient to the state. Its emphasis upon voluntary work and societal uplift declined with Gandhi's influence in the run-up to independence, and the party never seriously embraced a cadre-based structure premised on ideological clarity and indoctrination.

The flexibility of the Congress lay in its ability to represent broad subcontinental interests against the British within the framework of a secular party, under the control of a High Command. As the social basis of the Congress changed and broadened, so did the organisational principles of the party. At the time of the first non-cooperation movement, the Congress consisted of a president, three general secretaries, an All Indian Congress Committee (AICC), and provincial, district and 'taluka' (local) committees with various urban and municipal councils manned through regular party elections (Krishna 1966: 413–40).

Before the onset of parliamentary democracy and the rise of the parliamentary party, the Congress Working Committee (CWC) was the executive authority of the national organisation. By the late 1930s the CWC eventually came to consist of 20 members, 10 being appointed by the Congress president and 10 being elected by delegates from the AICC. It was responsible for organizing the meetings of the AICC. It was charged with the day-to-day running of party affairs and the interpretation of the party constitution. It directed and coordinated the Pradesh and district committees, having the power to act in the interests of the PCCs (provincial Congress committees) or the district Congress committees (DCCs) if necessary. If, on finding them unsatisfactory or unrepresentative, the CWC could dissolve and reconvene such committees accordingly.

The CWC was central to two party boards, the Congress Parliamentary Board (CPB) and the Congress Election Committee (CEC). The Parliamentary Board consisted of seven members, one being the president and one being (after independence) the prime minister as leader of the Congress parliamentary party. Five others were elected by the CWC. The Election Committee consisted of all members of the Parliamentary Board and seven other members elected from the AICC. The Parliamentary Board monitored the activities of the Congress assembly parties, while the Election Committee sanctioned the final selection of parliamentary candidate lists forwarded by the Pradesh election committees (Franda 1962: 248–60).

The reorganisation of the provincial Congress committees into 21 linguistically homogeneous units was another significant move that consolidated the Congress throughout India. With clearly defined financial responsibilities and locally accountable leaderships, these provincial committees quickly consolidated themselves around dominant social elites in the states that perceived the orders from the High Command through a prism of local and provincial issues. The basis of the Congress Party was thus neither exclusively regional nor ideological: it contained through the Hindu traditionalists links with members of the RSS, and the through the socialists, clear commitments to what would be by the 1950s, the policies of separate parties. Such parties would, however, seek to influence and lobby the Congress itself. Accommodation and compromise, determined in part by the various political levels of India, made the Congress into a loosely based structure that functioned under the auspices of a strong centre, aware that forceful or ideological issues would seriously divide the Congress and weaken its position vis-à-vis the British: 'the consensus within the Congress was defined in terms of certain broad social and political ideas...an essay in the art of the compromise' (Venkatasubbiah 1969). When serious social conflict did erupt 'the characteristic response was to try and arrange a compromise in the hope that divisive social and economic issues could be postponed for consideration, until after independence' (Frankel 1978: 30).

After independence, the apex party structures came into conflict with the Congress national parliamentary party and the authority of the prime minister. Issues of conflict lay much in over where power lay as well as over specific issues such as national languages, secularism and socio-economic policies that led to protracted conflicts between Hindu traditionalists and the more radical wing of the party under Nehru. Owing more to his personal charisma than to the office itself, Nehru asserted himself over Hindu traditionalists on national policy. Nehru was anxious to instill respect for India's Muslims and sensitivity to their particular position within Indian society after partition. This conflict overlay and to some extent animated the ongoing conflict between the party and governmental wings of the Congress. The categorical rejection of M. K. Gandhi's appeal to the parliamentary wing of the Congress to return to the party and commit themselves to social organisation after independence was already indicative of the tensions within the Congress, between state power, social uplift and rural idealism. Nehru's war of attrition against the election of Hindu traditionalist Tandon as party president resulted in Tandon's resignation and the assumption by Nehru of the offices of party president and prime minister until a more suitable – and compliant – candidate could be found. Differences over the approach to Hindu traditionalists (and their views on Hindi, cow protection, Hindu reform and Ayuvedic medicine as specific issues) also involved considerable tension between Nehru and his home minister, S. K. Patel.

Hindu nationalists, in the main, substitute organic social organisation for the mechanical institutions of the state in ways similar to Gandhian refutations of state action, and in the context of representative forms of government, they emphasised the long-term transformation over short-term strategies to mobilise votes (Hansen 1999; Jaffrelot 1996). The political project also has strong elements of corporatism, and conflict resolution through debate, mediation and then consensus. Not all have been hostile to participating within the state for power. The Hindu Mahasabha did have an overtly political project – indeed it was one of the first political parties to try and mobilise Hindu votes around the Babri Masjid at Ayodhya (in 1949–1950) – but its appeals to the symbolism of Lord Ram failed because of the dominance of the Congress, and in particular Nehru, who had so structured the national political environment to prevent them using religious appeals. The RSS was different in its conception and approach to politics. In some respects, it deliberately neglected the state, and with its resulting emphasis on social and cultural organisation – the Sangathanist networks, and reflective, almost contemplative context of the shakhas – it chose 'to work patiently on society over a long period, rather than seizing the state' (Jaffrelot 1996: 61). However, this lack of debate on the state should not be taken to literally imply a lack of interest, or a failure to recognise the powers of the state. Moreover, in response to state hostility by Nehru, and sustained attempts to thwart their political agenda after the death of Patel in particular, the RSS recognised the necessity of cooperating with the formation of a political party in October 1951 – the BJS – that closely mirrored their ideas (Graham 1990; Jaffrelot 1996).

In comparing *Hindutva* with forms of European fascism, Jaffrelot makes the interesting observation that, like Hitler, RSS writers (in particular Golwalker in *We or Our Nation Defined*) saw the state as a means, not an end, and – in a stark contrast to the Nehruvian view – believed that the state alone lacked the capacity to dominate (Kanungo 2002). Only through organising society could *Hindutva* come to dominate every aspect of life, including *eventually* the state itself (Eatwell, 1996). In the light of the experience of the 1990s, Khilnani has recently revised the view that the RSS is, in a critical sense, anti-statist: 'Hindu nationalist aspirations to redefine Indianess always presumed the availability of a strong state as the instrument through which to forge that identity' (Khilnani 1997: 190). When the RSS saw a possibility of gaining access to state power directly through their societal networks, it showed itself adept at setting up or infiltrating political parties: in foul weather, it would revert to its earlier social welfarism and sit out periods of persecution or reaction, but the long-term project required a grounding through society itself, eventual access to the state, and at the least, the indifference of the state to its societal project.

The vision of Hindu sangathan set out by V. D. Savarkar and implemented by K. B. Hedgewar (Zavos 2000: 167–210) was actually a direct inversion of Nehru's vision; it started from within society and moved upwards, while Nehru's obsession with the rational modernism of the state and the irrationality of society compelled him to move downwards. Both visions of the social order were revisionist. Nehru sought to abolish caste and the RSS sought to educate lower caste identities into the fold of Brahmin civilisation using an organic model of society as a harmonious structure, close in this respect to Gandhi's praise for a pure and unadulterated form of the Jajmani system. The RSS could live quite happily with the Indian state, disagreeing with Nehru's multiculturalism, confident in the view that it would eventually mirror a society dominated by *Hindutva* identity. In redefining Hinduism into a form of cultural nationalism, *Hindutva* was largely immunised against the secular hegemony of Nehru and the ambiguities of the Indian constitution, although it was to remain vulnerable to specific banning – such as carried out in the wake of Gandhi's assassination – and specific legislation aimed at restricting its practice and use of specific symbolism. Thus ironically, despite the differences in emphasis, the two ideas of India 'operated within the same set of discursive tools and the same range of political strategies' (Zavos 2000: 216; Khilnani 1997: 190).

Moreover, the idea and imagery of the RSS as a sort of invisible social organisation gave it a ubiquitous presence. Linked to networks of educational and welfare institutions, and drawing upon the selfless commitment of local shakhas and the dedicated *pracharak's*, the RSS drew upon quite a different idea of modern organisation. The image of selflessness, of the saintly idiom of the *pracharak* as the 'new Hindu', clearly attracted a large number of middle class Brahmin men, concerned over perceived Congress appeasement to Muslims and, more indirectly, concerns over the consequences of mass politics which exposed high caste Hindus to lower caste mobilisation (Jaffrelot 1996: 71). The true numbers of RSS volunteers was a source of constant anxiety for Nehru, as was his recognition that sympathies for their dedication shaded off into Hindu traditionalists and many Congress members. Jaffrelot notes that Nehru's decision to ban it in the wake of Gandhi's murder lay in his firm conviction that it was a conspiratorial force, directly comparable with fascism, and that the Mahatma's death was just the start of a campaign of terror, a view that was widely held. As an editorial reported 'the case of the RSS is one of the mysteries of modern India but a mystery which ought not to be tolerated. It seems to embody Hinduism in a Nazi form and whatever be its professional appearances there [is] something secret behind its physical parades and that secret is becoming more sinister' (Graham 1990: 11; *The National Herald* 1948). Significantly, Mrs Gandhi would come to share these fears in the mid-1970s, when in a slightly changed landscape, she saw the RSS and the BJS party as the principle force against her and her own socialist radicalism.

Significantly, S. P. Mookerjee, a member of the Hindu Mansabha and of Nehru's interim cabinet, initiated the formation of the BJS. He launched the new party after consultations with the RSS drew at first an indifferent response, but later, the organisation responded in effect by co-opting its organisation and moulding it, with intermittent success, along the same ideological lines as the RSS itself. Provoked in part by the need to lobby more effectively against state oppression as witnessed in the wake of Gandhi's death, and also through the need to more directly challenge the Nehruvian agenda in the wake of Patel's sudden death in 1950, the RSS provided the manpower and mobilising spine of the BJS. Its emphasis on social networks and welfarist agendas conflicted with Mookerjee's belief that the party should, in the main, mirror the political structures and electoral strategies of the Congress (Graham 1990). Tensions over organisation and electoral strategy characterised the relationship between the BJS and the RSS from the start, with the RSS favouring secret debates and then, in the wake of a consensus, acceptance of the party over the individual leaders. Hierarchical in nature, and somewhat opaquely but directly linked to the command structure of the RSS, Jaffrelot notes that by the 1950s the BJS faced a serious paradox: its electoral appeals to a Hindu ethno-religious nationalism would only work once the Hindu majority had been transformed to see themselves as an ethnic nation and demand the state recognise them as such, but in order to facilitate this, and in effect, to remove the wider constraints placed upon it by the Congress dominance of the state, it needed to engineer societal identities over the long term. Winning elections at the expense of its ideological purity, either through populism or coalition, would create strains between electoral and societal strategies, and the link between the RSS and the BJP would, until the late 1980s, engender suspicion on behalf of self-professed secular political parties who otherwise would have cooperated to combine against the Congress.

Secularism was not explicitly mentioned in the Indian constitution until the 42nd Amendment of 1976, although an attempt had been made to do so in 1949. In principle (but not in practice) secularism in India equals 'public respect for all religions, rather than the state promotion of a public culture opposed to, or sceptical of religion' (Harriss-White 2003: 136). During the Constituent Assembly debates, Professor K. T. Shah sought to introduce an Article setting up a specific 'wall of separation' that explicitly prohibited the state from interfering in matters of religion, strictly confining it to individual rights. It was rejected on the grounds that it clashed with the need for the state to define, in many cases, the boundaries between religion and culture in India, and more pertinently, to interfere in matters of religious practice on the basis of social reform, especially where such practice was deemed to be socio-economically backward or reactionary.

This mode of interaction was to problematise the practice of secularism from the beginning:

> the mode in which the judiciary has attempted to police the boundaries between the secular and the religious has been to engage in a process of defining elements of the religious itself. The nature of the emerging secular order is dependent upon prevalent conceptions of religion, and the reformulation of religion is powerfully affected by secular institutions and ideas.
>
> (Galanter 1998: 268)

However, ensuring the fundamental rights to equal treatment of men and women, for example (through divorce or polygamy), or equal access to temples, or by reforming unacceptable social practices such as untouchability or child marriage would require the state to intervene into what could be rightly defined as a religious context (Harriss-White 2003). As Tambiah has noted, however, 'it is not clear on the conditions under which a general conception of fundamental rights applicable to all citizens supersedes those religious pratices that are deemed discordant with those rights?' (Tambiah 1998: 425). This was one of the most complex difficulties facing India's modernising elite at the time of independence, an implicit clash between the fundamental rights of the citizen and a commitment by the state for broad-based social uplift (involving the need to identify and act on behalf of disadvantaged communities) in order to implement a socialist pattern of society.

The demand for a bill of rights stemmed from the nature of the Congress' struggle with the British, and the influence of European concepts of natural rights and the Congress' own experience of the coercive aspects of state action. Congress demands for a charter or bill of rights were relatively longstanding. At the time of the 1935 Act, attempts to lobby the British for their inclusion were condemned by Lord Simon on the grounds that it was not possible, in the current conditions pertaining to India, to invest rights in individuals as opposed to communities, and even then, such guarantees were effectively useless without a means to implement them. Simon argued that it was pointless granting rights that the state could not always guarantee and which, under due process and amendment, it could actually remove or restrict as seemed fitting. The Report of the Simon Committee of the Indian constitution noted in 1930 that:

> many of those who came before us have urged that the Indian constitution should contain definite guarantees for the rights of individuals in respect to their religion and a declaration of the equal rights of all citizens… Experience, however, has not shown them to be of any great practical value. Abstract declarations are useless, unless their exists the will and the means to make them effective.
>
> (Simpson 2001: 17)

Yet in the constitution enacted on the 26 January 1950, Congress effectively rejected this quintessentially English conception of 'natural, common' law for a more continental understanding of codified rights (Simpson 2001: 213). The bill of rights occupied Part III of the constitution, Articles 12–33, and, as will be discussed, was referred to extensively in Emergency Articles 358 and 359. Article 13 set out to protect the fundamental rights from executive amendment by granting an independent judiciary powers of review over Parliament. Article 13(ii) stated that 'moreover, any law that is made that takes away or seeks to abridge the rights guaranteed in Part III of the constitution shall be declared null and void' (CAD 1947: 423). Articles 14–18 invoked rights of equality, such as the abolition of untouchability and other forms of social discrimination. Article 19, the keystone to the entire chapter on fundamental rights, contained an important list of freedoms for the individual, stating that 'all Citizens shall have the right to (i) freedom of speech and expression, (ii) to assemble peaceably and without arms, (iii) to form associations and unions, (iv) to move freely throughout the Indian Union and (v) to settle anywhere within the Indian Union, (vi) to require, hold and dispose of property and (vii) to practise any profession or to carry out any occupation or business' (Basu 1961: 483).

Articles 20–22 protected the citizen against the loss of life or liberty from conviction or detention 'except by procedure established by law', while

Articles 23–30 crucially related to specific rights of culture, religion and education aimed at defending minority rights. Significantly, with reference to religious freedoms, Article 25 set out the provision that all persons were equally entitled to freedom of conscience and the right to freely profess, practice and propagate religion (even to change their religion through voluntary conversion) subject to public order, morality and health, and the states right to regulate economic and financial matters deemed to be of secular concern. These same conditions – to be identified by the state itself through legislation and judicial review – were also attached to the recognition of the right of religious denominations to establish and maintain their educational and social welfare institutions, including customary personal laws relating to divorce, marriage, adoption and inheritance.

The matter of identifying the conditions and the rights involved when seeking to balance individual claims against group rights was made even more complex by the enactment of a series of directive principles, which were neither compulsory nor legally enforceable, but which were nonetheless placed into the constitution. When discussing the Supplementary Report on Fundamental Rights, Vallabhbhai Patel stated in the Constituent Assembly that 'we have come to the conclusion that, in addition to justiciable [*sic*] fundamental rights, the constitution should include several directives to state policy, which though not cognisable in any Court of Law, should be regarded as fundamental to the governing of the country' (CAD 1947: 423).

Consequently Articles 37–51 and 55 of the constitution express the commitment of the state to promote public welfare, to prevent the concentration of ownership and ensure equitable distribution of essential commodities and the right to be gainfully employed. Given the fundamental inequalities throughout India at the time of independence, both with regard to the social hierarchy and land ownership, the directive principles spelled out nothing short of a social revolution, to eliminate 'medievalism based on birth, *religion*, custom and community and reconstruct her social structure on [the] modern foundations of law, individual merit, and secular education' (K. Santhanam cited in Austin 2003: 69, emphasis added). The directive principles contained the caveat that such policies would be undertaken 'within the limits of the State's economic capacity and development'.

Commenting on this section of the constitution, however, Dr. Ambedkar said 'In my judgment, the directive principles have a great value, for they lay down their own idea of economic democracy and social justice' (Jhabvala, 1980: 82). One of the most controversial aspects of the directive principles was to be not just the commitment of future governments to enact a basic socialist pattern of society, but to legislate for a uniform civil code – Article 44. Putting aside the British traditional practice of treating communities through their own personal laws with reference to the age of consent, divorce and inheritance, a uniform civil code aimed to create a common basis for treating these matters and in doing so, strengthening a sense of civic nationalism above, and superior to, various religious codes enforced by a variety of differing authorities.

Despite the recognition of 'differing faiths' implicit in Article 44 (and the implicit neutrality of the state with regard to differing religious communities), the position of the Hindu traditionalists towards the uniform civil code (and later that of the BJP in the 1990s) was to be less hostile than their criticism over the Congress's evident reluctance to legislate for it. State intervention into personal law, through which the state entered into religion to change not issues of belief but, it argued, matters of social practice, was, in the context of India's religious pluralism, concentrated disproportionately against Hinduism from the beginning (Smith in Bhargava 1998: 177–233). In reality, the state engaged unequally with different religions, and created the general impression that, in allowing Muslims to decide their own pace, the party could represent and speak on behalf of the Hindu majority (Harriss-White 2003: 168; Larson 1995: 178–277).

In 1948, Congress introduced The Hindu Code Bill, which sought to modernise Hindu personal law. Withdrawn and later passed in the guise of three specialist bills, the legislation left minority religious communities alone, despite the commitment to a uniform civil code and the recognition that 'a secular law, equally applicable to all citizens of India, will be a notable achievement in the process of establishing the secular state'. Conscious of their potential vulnerability as a minority, and as a potential hostage to the vulnerabilities of partition (and the accusation of divided loyalties), Nehru deliberately sought to reform the majoritarian community first, and as such – given the failure to enact a uniform civil code by any subsequent government of India – was seen to act unevenly and discriminately.

While Nehru was keen on reforming or mitigating certain abuses in Hindu society, he at the same time maintained that Congress should be generous to the Muslims, and desist from intrusively disturbing Muslim sensibilities by advocating reforms to their practice that appeared undemocratic or inegalitarian (Tambiah 1998: 424). By the mid-1980s, in the context of the Shah Bano case, this would become one of the most serious charges against Congress; pseudo-secularism, and that it had overtly sided with traditionalist Muslims who were not interested in reform nor, specifically, monogamy (Larson 1995: 256–60).

There were other ways in which the state, unintentionally, both agitated and strengthened religious sensitivities. One was through the policy of affirmative action, which sought to correct historic social and economic inequalities. In spite of the intentions of the secularists within the Constituent Assembly, the consequences of political action carried out by the state were to be radically different than otherwise intended. Indian society had developed a close association between religious identity and economic and social wealth both within the caste system, and within the Muslim community that remained behind in India after partition. Compared to the Mahajjirs, Indian Muslims were predominantly poor. In 1980, Muslims were half as likely than Hindus to be below the poverty line, and Muslim illiteracy was approximately 15 per cent higher than Hindus (Harriss-White 2003: 166). Within the Hindu community, caste and jati, fragmented through an intimate association between ritualised polluting status and occupation, meant that the poor largely corresponded to low caste shudras and recognised scheduled castes and tribes (Galanter in Bhargava 1998). Thus stressing economic entitlements over confessional identity made little difference in practice, since it appeared to all intents and purposes to be using the same signifier. Indian social policy, while seeking to outlaw caste and the overt use of religion for social reform, ended up using it to identify socio-economic groups that required socio-economic assistance from the state: thus 'paradoxically, it is because of, not in spite, the secularist ambitions of the constitution that distributive policies are organised in part around religions.' (Harriss-White 2003: 136). Moreover, the consistent widening of affirmative action, especially in the context of the OBCs, which constituted over 54 per cent of the Indian population by the early 1990s, generated concerns among the forward castes that they would lose out and jeopardise their material – and sacred – status.

In dealing with the problems of the minorities throughout India, the Advisory Committee on Minority Rights to the Constituent Assembly categorically rejected the British practice of awarding communal electorates as divisive, although it recognised multi-seat constituencies until 1961 and a restrictive policy of limiting candidates to specific social categories under reservation in specific constituencies. Reservations were formally laid down within the constitution for various sections of the population defined on the basis of socio-economic backwardness. Article 15(i) introduced to scheduled castes (harijans or, later still and more subjectively, dalits), scheduled tribes and OBCs the principle of 'protective' or proactive discrimination (Wadhwa 1975). The 5th Schedule of the constitution noted that 'the constitution has directed the State to make special provisions for any socially and educationally backward classes in addition to the scheduled castes and the scheduled tribes'.

Smith in his classic treatment on protective discrimination noted that it was prevalent in three main areas: quotas to public employment, access to state education and issues of voting or representation. The identification of groups claiming these identities (and hence the entitlement of affirmative action) were to be decided on by a series of state commissions and a series of ad hoc reports by a national Commissioner for Scheduled Castes and Tribes (Smith 1968: 16). During the 1980s, the electoral and political significance of the OBCs would be a critical fault line with reference to political mobilisation during elections and for governmentality generally. Again, the paradox of recognising equal citizenship, while also clarifying the basis on which discrimination by the state could be carried out, was to complicate the Nehruvian modernisation plan and the state's institutional capacity to negotiate and implement specific strategies of provision within a democratic polity. Involving as it did differing institutions championing differing rights – initially Parliament for group rights and the judiciary for individual rights – this struggle took place within government as much as without.

These complexities of Indian identity, and the mission of the state to modernise society, were carried out in an institutional context that was, on the one hand, an embodiment of the Westminster-style system of government inherited from the British, and on the other, a written, elaborate, easily amendable constitution resting upon the interpretative powers of a supreme court. Alongside the paradox of a secular polity that drew upon religious identities to redistribute societal power, there came to exist an implicit contradiction between parliamentary sovereignty (and the supremacy of party-based government within the executive) and an unelected court able to strike down parliamentary legalisation on the grounds that it was *ultra vires* of a written constitution. The fault lines of this conflict would be centre–state relations and the nature of Indian federalism, socio-economic reform and redistribution, and eventually the role and nature of *Hindutva* and the BJP.

This need for a clearly central federalism was enforced by the belief that it would provide the most conducive institutional framework for socio-economic advance. In a report of the Union Powers Committee in 1947, Nehru declared that

> now that Partition is a settled fact, we are unanimously of the opinion that it would be injurious to the interests of the Country to provide for a weak Central authority, which would be incapable of providing peace. At the same time, we are quite clear in our minds that there are matters in which authority must lie solely with the Units... To frame a constitution on the basis of a unitary State would be a retrograde step.
>
> (CAD 1947: 60)

However, the Union Constitution Committee, despite having decided upon three exhaustive lists of legislation allowing for provincial (state), concurrent and central policy, nonetheless invested power in the centre. Even apart from the Emergency acts, the recognition of differing levels of legislative competence was premised on the powers that the centre would take back what it had devolved. Nehru commented 'In the matter of distributing powers between the centre and the units, we think that the most satisfactory arrangement is to draw up three exhaustive lists on the lines followed by the Government of India Act 1935' (CAD 1947: 62).

When the lists were presented to the Constituent Assembly on 20 August 1947, the central list contained 87 subjects, the state lists 57 entries, and the concurrent list contained 36 entries (Pal 1984: 78). The concurrent list would automatically fall into the ambit of the centre where state legislation contradicted central legislation. Article 229 allowed the centre to legislate on the state subjects following the consent of a particular state.

Financial arrangements, including provisions for grants-in-aid and the rights of taxation (with the significant exception of agricultural taxes), were heavily biased to the centre, resulting in the creation of a powerful Finance Commission (set up to review arrangements every five years) to prescribe, oversee and generally adjudicate over the division of monies between the centre and the states. Article 263 sanctioned the setting up of an inter-state council if at some future date such an institution was considered necessary to resolve anomalies within the federal system.

Yet the centralism within the Indian system was quite plain for all to see. The national parliament even had the right to alter and divide state boundaries in potential disregard of the states in question. Tellingly, Dr. Ambedkar defended the implicit unitary state on the grounds of modernisation: 'Some critics have said that the Centre is too strong...The Draft constitution has struck a balance. However much you may deny power to the Centre, it is difficult to prevent the Centre from becoming strong. Conditions in the modern world are such that the centralisation of power is unavoidable' (CAD 1948: 42). It was argued by several independents within the assembly that the scheme adopted by the Constituent Assembly involved the worst traditions of 'nominal' federalism adopted under the British.

With reference to the executive, the Constituent Assembly adopted in outline the basic framework of the British cabinet system. This involved the Union president acting as the republican substitute for the British monarch. In this capacity the Union president acted as head of state but not head of the executive, the latter being the prime minister. Since the powers of the Union president were considered to be titular, it was decided that they should not be elected on a mass franchise, where their influence might come to be greater than the prime minister (who would only be elected from a single member constituency). Articles 52–61 dealt with the election of the Union president, Articles 62–71 with the elections of the vice president and Articles 74–78 with the establishment of a council of ministers under a prime minister. Both the council of ministers and the prime minister would be drawn from a popularly elected lower house (Lok Sabha) on a first past the post electoral system. Elections to the Lok Sabha would be held every five years. It was decided that the president would be elected on the basis of a complicated electoral college drawn from the state legislatures and the Lok Sabha through a variation of a first past the post electoral system incorporating a single transferable vote (Quraishi 1973: 32).

The Rajya Sabha – the upper house – was to be excluded from this electoral college, since it already incorporated an element of indirect election and nomination (Quraishi 1973: 28). The electoral process consisted of a continual ballot aimed at ensuring that the successful presidential candidate achieved a prescribed majority of over 50 per cent. If such a majority was not achieved on the first count, the lowest candidate would be dropped with the votes going to the candidate indicated in the second preferences of that candidate's voters. Such a procedure was adopted to ensure the representative nature of the president (and prevent marginal victories) and to give various minority candidates the chance to obtain a majority through the construction of alliances based upon the second preference vote (Quraishi 1973: 32). The presidential electoral college gave large populous states such as Uttar Pradesh a significant position in presidential elections. Nehru stressed the imperative of keeping the presidential electorate separate from the central Union legislative: 'the central legislative may and probably will be dominated by a party or group that will form the Ministry. If that group elects the President, the President and the Ministry will represent exactly the same thing' (CAD 1947: 735).

Of vital significance to the future of the constitution was the exact relationship between the president as head of state and the prime minister as head of the executive. All action would be carried out by the prime minister, advised by the council of ministers, and implemented in the name of the president. The relationship was defined within the constitution by Articles (74, 75, 78 and the Third Schedule): 'The Prime Minister and the Council of Ministers shall aid and advise the President.'

The president would invite the leader of the largest party in the Lok Sabha to become prime minister and form the government under Article 71(ii), while Article 75(ii) gave the president the corresponding right to remove the prime minister once it was clear that he or she no longer commanded the leadership of the largest party in Parliament or if the party had itself lost a majority. Article 143 also gave the president the right to refer any legislative matter from the national government, or even bills pending from states, to the Supreme Court if, in the president's opinion, it contained provisions that might be in contradiction with the constitution. Such presidential discretion also extended to legislation passed by state governments as well, and was to be a moot point with non-state Congress governments from the 1970s onwards.

The relationship between the prime minister and the president was not exhaustively debated at the time of independence. It was the opinion of the Constituent Assembly that the relationship would settle better on the basis of establishing conventions as opposed to laying down strict constitutional regulations. It was never clear, for example, whether the advice tendered to a president from a prime minister would be binding. Article 78 stated that the prime minister must 'communicate to the president all the decisions of the Council of Ministers relating to the administration of affairs and proposals of the Union' (Jhabvala 1980: 109). It was recognised, however, that through this link between the head of the executive and the head of state many other institutions and designated constitutional procedures intersected state power: the role of the governors, the necessity and scope of judicial interpretation, and more specifically, special provisions of governance, such as ordinance and emergency legislation. Article 123 empowered the president, in the absence of a sitting Parliament, to pass urgent legislation. The Article read that 'if at any time the federal Parliament is not in session and the president is satisfied that circumstances exist which render it necessary for him [sic] to take immediate action he may promulgate such ordinances as the circumstances appear to require' (CAD 1947: 748).

Such ordinances would have to be passed by Parliament before automatically lapsing after three months. Although the promulgation of ordinances was again tied into the conventions of prime ministerial advice, such powers, inherited from the 1935 Act, seemed to some to create the possibility of a presidential dictatorship. H. V. Kamath observed '[T]he constitution provides the President a Council of Ministers to aid and advise him ... but there is no injunction laid upon him to accept that advice' (CAD 1949: 106). Dr Ambedkar had earlier reassured the Constituent Assembly by stating that 'the president of the Indian Union would be generally bound by the advice of the ministers ... He can do nothing contrary to that advice so long as his Ministers command a majority in Parliament' (CAD 1948: 32).

The granting of ordinance powers to the executive was seen as further abridging the federal principles of the constitution, as was the presence of a nominated – as opposed to an elected – governor. The executive of the state units (Articles 152–167) involved the nomination of governors by the president to aid and advise chief ministers, elected by Members of the Legislative Assembly (MLA), who were in turn popularly elected on a constituency basis. The relationship between the chief minister and the governor was parallel to that envisaged between the prime minister and the president. The chief minister was invited to form a government by the governor on the basis that he controlled the largest number of seats (i.e., the largest party) in the state assembly. Powers to issue ordinances were granted to the governor under the auspices of the president, himself acting under the council of ministers.

The concept of nominating state governors was controversial from the beginning. Dr Ambedkar said that nomination would ensure the selection of elder statesmen and 'experts' as opposed to party men and hacks. The powers of the governor would not checkmate the powers of the chief minister unless the chief minister lost the confidence of the state assembly, or the chief minister sought to act in an unconstitutional manner. Yet opposition members argued presciently in the Constituent Assembly that 'suppose a party, which is hostile to the party in power at the centre, comes to power in a province [state]. And suppose there is a quarrel between the provincial government and the central government. Immediately the president, thinking that there is domestic violence, can suspend that part of the constitution according to emergency law' (CAD 1948: 936).

Ali Sahib Bahadur condemned the Union arrangements on the grounds that 'the provinces will be nothing but glorified district boards' (CAD 1948: 296). The governors were to be the 'prime minister's men' (and women), constrained only through presidential discretion and the conventions that controlled prime ministerial ambition.

Various anxieties over the role of a nominated governor and the possible route towards centrally backed authoritarianism came together in opposition to Article 356 or President's Rule. This Article, which was indeed to become the most abused Article within the constitution, stated that 'if the President, on receipt of a report from a Governor or otherwise', was informed that the constitutional machinery of a state had broken down he could bring the state under central legislation. Article 357 transferred the powers of legislation from the state under suspension to the Union. The proclamation, made in the name of the president, would automatically lapse in six months if not passed by the Lok Sabha. Ambedkar defended the generality of the Article by stating that

> it may be that the Governor doesn't make a report…we must give the liberty to the President to act even when there is no report by the Governor and when the President has got certain facts within his knowledge on which he thinks he ought to act to fulfil his duty.
>
> (CAD 1949: 135)

President's Rule, when declared, would suspend state assemblies in order to organise a fresh administration, or dissolve an assembly and order fresh elections when this was no longer possible. Decisions concerning suspension and dissolution were envisaged in the light of the governor's report or in the light of information available to the president. This Article was widely condemned on the grounds that it allowed for arbitrary action on behalf of the president not in keeping with his other titular functions. Oddly few in the Constituent Assembly recognised that it would actually allow a prime minister, through presidential power, the possibility of removing state governments hostile to the national government.

In the context of partition and the quite palpable fears of national disintegration, there was much fear of an excess of centralism leading to authoritarianism. In the complex meshing of a written constitution premised on parliamentary government, it was not clear who the Caesar might be – the president or the prime minister. Moreover, by the mid-1950s, there were also concerns that the powers acquired by the courts might frustrate popular government, especially given their involvement in upholding fundamental rights and on deciding the context in which affirmative and redistributive policies could be carried out pace Article 15(I). Articles 124–147 dealt in detail with the setting up of the Supreme Court as well as the relationship envisaged between the Supreme Court and the high and district courts. Unlike the legislative and executive authority, the court system in India was unitary with the highest court being the Supreme Court. From the outset the Supreme Court was given the rights of exclusive jurisdiction of interstate disputes, jurisdiction on matters arising from treaties entered into by the Union, matters concerning areas allotted to the centre and referred to the Court for advice, powers involving substantial issues of constitutional interpretation and appellate powers for purposes of defending the fundamental rights. Within the institutions of the state, and despite explicit debates within the Constituent Assembly to avoid excessive litigation by substituting the phrase 'process established by law' for the more anglophile 'due process', the courts were to emerge as powerful actors of governance in their own right. In the Constituent Assembly debates many, like Pandit Thakar Bas Bhargava, saw the courts as the key defender of liberty, because 'even if the legislature is carried away with party spirit and is sometimes panicky, the Judiciary will save us from the tyranny of the legislature and the executive' (CAD 1948: 844).

Whatever the term used to describe the role of the Court, it was clear from the onset that, as an institution, it would acquire major significance as a key interpreter of the scope and direction of government policy. In the final analysis, the courts were the defenders of the much-cherished bill of fundamental rights. The whole viability of Part III of the constitution rested upon Articles 32 and 33 that guaranteed the right of an Indian citizen to move the Supreme Court for the enforcement of his or her fundamental rights. This in turn rested on the scope and ability of the Court system to interpret and enforce constitutional provisions. However there remained within the Indian legal tradition the ambiguity of a constitutional scheme that set out a bill of rights, and the common law practice of habeas corpus that was deemed to be independent of the states right to grant or deny substantive liberties. Despite the unitary structure of the Indian legal system, the high courts would invariably refer to legal precedent outside of a strictly codified bill of rights, however amended by the constitution. This was to have profound consequences for matters such as land reform and socialist reform generally (see Chapter 3), and for periods of Emergency in which the state sought to free itself from the constraints of judicial intervention by the extraordinary Acts of 352 and 359 (Austin 1999).

At the time of the Constituent Assembly debates, the issue of fundamental rights and the powers of government to compromise them were much discussed. Apart from the issue of setting individual rights aside in the context of group rights and affirmative action, there was the wider debate of providing the state with security to defend itself in times of crisis. Again, it is critical to recall that these debates were taking place in the context of unprecedented communal violence and the movements of populations. The survival of the state was uppermost in the minds of the Constituent Assembly. In a letter to Dr. Rau, constitutional advisor to the Assembly, Shri Ayyar remarked that 'recent happenings in different parts of India have convinced me more than ever that the fundamental rights guaranteed under the constitution must be subject to public order, security and safety, though such a provision may to an extent neutralise these rights guaranteed under the constitution' (Austin 1999: 70).

In uncanny conformity to Lord Simon's observations back in the late 1920s, the Articles defining rights went on to specify the context in which they would be curtailed. After listing salient rights, Article 19 went on to state that 'nothing in subclause (a) of section (i) of Article 19 shall effect the operation of any existing laws in so far as it relates to or prevents the State from making any laws relating to slander, defamation, contempt of court or any matter which offends against decency, morality or tends to overthrow the State' (Basu 1961: 483). Pandit H. N. Kunzru argued that the fundamental rights were so surrounded by legal restrictions as to be, in reality, legally unenforceable. Somnath Lahiri made the observation that 'almost every Article is followed by a proviso that takes away that right almost completely...they constitute fundamental rights from the view point of a police constable (CAD 1947: 384).

Furthermore the Indian constitution seemed to anticipate the future enactment of a preventive detention bill by outlining the rules and regulations that would have to be followed if such a bill was ever placed on the statute books. Article 22(4) stated that no laws providing for preventive detention shall authorise detention of a person for a period longer than three months, unless advisory boards have reported before the expiring of the initial three-month period that such detention should be continued.

The first preventive detention Act was introduced in 1950 by Sardar Patel, Congress home minister, with particular references to the Telengana peasant rebellion in Andhra Pradesh. The life of the Act was renewed on six separate occasions and amended eight times prior to 1969 alone. It was allowed to lapse for a few years after 1969 for political considerations but was duly re-enacted in 1971 after the Fifth Lok Sabha elections (Bayley 1962: 1). Bayley remarked that 'it was not the fortune of the Preventive Detention Act to be enacted, to serve its allotted time, and quietly fade...it was found useful and continued, found wanting in parts and was modified' (Bayley 1962: 17).

Durga Das Basu begins his eight volumes of commentary on the Indian constitution by stating that almost 75 per cent of the 1950 constitution owes itself directly to the 1935 Government of India Act. Articles 352–360 and 365 were taken from or inspired by the 1935 Act, and were a direct consequence of the fears evoked by Partition for the future of the Indian Union. Article 352 allowed the president to declare a State of Emergency throughout India on the grounds either of external aggression, internal disturbances or a financial crisis. Article 352(ii) states that 'a proclamation of Emergency, declaring that the security of India or any part of the territory thereof is threatened by war, external aggression or internal disturbances may be made before the actual occurrence if the President is satisfied that there is imminent danger thereof' (Basu 1961, vol 1: 4). The courts were explicitly barred from being able to examine presidential satisfaction and could not deliberate upon whether that satisfaction contradicted any of the terms within the constitution. It was envisaged that the emergency powers of the president would again be governed under prime ministerial advice on the lines controlling ordinance powers generally.

Once the proclamation was issued, the rest of the emergency clauses would come into power automatically. These included Article 353, which immediately empowered the centre to abridge the Seventh Schedule. Article 353 also extended the powers of the executive and allowed the president to extend the life of the Lok Sabha by one year at a time by a simple majority vote within the Lok Sabha itself. Article 354 allowed the centre to make allocations within centre–state financial arrangements, supplementing Articles 268–279 that allowed the president to act concerning funding generally. Article 355 codified the duties of the Union concerning the defence of the individual states to ensure that the government of every state is carried out in accordance with the provisions of the constitution.

Most dramatically, Articles 358 and 359, which automatically came into effect under Article 352, suspended the provisions stated by Article 19 concerning the fundamental rights. Article 359 suspended the right of the Supreme Court to defend any other fundamental right and ensured the technicality that it was not the rights as such that were suspended but the right to act in their defence. Both these Articles ensured that the citizen should have no protection against the executive authorities during a proclamation of emergency. Thus the executive, subject to a simple two-thirds majority vote in the legislature, was empowered to act in complete disregard of Article 13(ii). Basu has commented that 'the peculiarities of these Emergency provisions... relating to the suspension of our fundamental rights is that no distinction is herein made between times of war and peace' (Basu 1961: 191).

It was this extraordinary measure that provoked concern within the Constituent Assembly. K. T. Shah warned the Assembly 'the danger of substituting such a thing as "internal disturbances" for "violence" is very serious because internal disturbances can be defined according to the mood of the moment' (CAD 1949: 112). Z. H. Lari criticised the Emergency Articles on the grounds that it preserved the worst aspects of the 1935 Act, especially with reference to Article 359. Pocker Sahib Bahadur stated that 'the powers of the Courts should not be made to depend upon the will and pleasure of the government and they should under no circumstances be allowed to interfere with the powers that rest in a Court of law' (CAD 1948: 943).

The 1950 constitution concluded with Articles dealing with constitutional amendment. Jhabvala comments that the Constituent Assembly, in adopting one of the largest constitutions in the world, wanted to minimise the risks related to excessive legalism and rigidity in conditions of rapid social change. It was thus decided to make the constitution flexible to the demands of a new country. Three provisions were adopted concerning constitutional amendment in relation to specific parts of the constitution. The first concerned simple majority voting in the Lok Sabha. The second provision required a two-thirds majority vote in both houses of Parliament. The third provision required the passing of a two-thirds majority and the ratification of the bill in half of the states. The third and most extensive process of ratification concerned 'serious' or 'structural' changes to the constitution such as changes in the presidential electoral, or changes to the divisions of powers between the centre and the states. The second procedure determined the form of a majority of constitutional amendments. It was to be of great significance later that the Constituent Assembly had seemed to extend the flexibility of constitutional amendment to the fundamental rights themselves, and had, in the context of a written constitution, obscured the sovereignty of parliamentary doctrine that the British so treasured.

On the whole the constitution was perceived to be a Western import, indigenised and modified around the perceptions of a westernised political elite and applied to a rapidly changing and diverse political context. Its flexibility rested on a series of complex interrelated institutions, and an idea of India that was broad stroked and to some extent negotiable. But it was also ambiguous. The socialist pattern of society was closely imbricated with socio-religious and cultural signifiers and conflicting ideologies as to who or what constituted the nation. More significantly, the key institution was a political party, which contained aspects of both ideologies. The absence within the Constituent Assembly of a powerful Hindu nationalist contingent should not detract attention away from the degree of that ambiguity, or its potential for structuring political institutions. Hindu traditionalists, although lacking the coherence and ideological depth of Hindu nationalists, none the less kept up within Congress a constant pressure on single issues such as Hindi, cow slaughter, aspects of Ayuvedic medicine as well as a more sustained opposition to the Hind Reform bills. Throughout northern India, at the local level, traditionalists were far more powerful than their presence within the national Parliament implied. Nor is it clear that, apart from specific issues on minority rights, and the emphasis on a multi-religious, multicultural India in such areas as education, welfare and individual rights, the emergence of the Indian state after 1950 was radically different from that envisaged by the RSS. Given the societal emphasis and its mode of operation, it is difficult to be categorical on this. Did the RSS envisage a non-democratic, authoritarian state? Or did it merely image a homogenous Hindu India in which cultural impurities had been removed or assimilated, an organicist vision of society that would have supported a state operating through a guided consensus (as witnessed throughout the RSS itself) avoiding open debate and dissension? Would it have been more centralist than federalist? (It is hard, in abstract, to imagine a more centralised state.) Would there have been more emphasis upon localism, on a sort of Gandhian Panchayat Raj that, in keeping with the traditions of a pure Jajmani system, organised a fundamentally different mode of production than that encouraged by Nehruvian socialism and the mixed economy? At the time of independence, the RSS coexisted with, and adapted to, the constitutional settlements of India. And it was willing to coexist with, and indirectly participate within, democracy itself.

3 Political mobilisations 1963–1971

Chapter 2 discussed the eclectic nature of Congress ideology, and the relationship between the party structures and those of the Indian state as set out in the 1950 Constitution. A majority of these structures dated back to the early part of the twentieth century, and although clearly colonial in origin, the large degree of overlap between the 1950 Indian Constitution and the Government of India Act 1935 reflected both the familiarity of India's political elites with this particular format of governmentality as well as a continuity of concerns: a fear of instability and chaos resulting from the onset of independence, of insurrection and territorial disintegration.

How did this complex institutional amalgam between party/government and state work throughout an uneven and elite-based society? In what ways could the resulting totality meaningfully be called a democratic polity? Many commentators noted a paradox in the adoption by a predominantly agrarian society – characterised by high levels of ritualistic and socio-economic inequalities – of a democratic system premised on one citizen, one vote. Other commentators, most notably Barrington Moore, noted the contradictions of a predominantly rural, peasant-based society adopting a democratic constitution, better suited to the urban-based working and middle classes (Barrington Moore 1993). Others referred to the absence of a formal party system and the probability that Congress dominance would persist for some time to come. Many of these observations were in the main correct, and although they risk ignoring the experience of democracy and its ability to act as a solvent on structural inequalities, India is probably best understood at the time of independence as a liberal oligarchy (Brown 1985; J. A. Gallagher *et al.* 1973; Sarkar 1983; Tomlinson 1993).

W. H. Morris-Jones once observed that 'the way in which parliamentary democracy works depends, more than we might like to admit, on the balance of powers between political parties' (Morris-Jones 1957: 113). Of all the concerns expressed within the Constituent Assembly, the proximity of the ruling party to the state, to the initial exclusion of all opposition parties, had been a continuous theme. On 7 November 1948, Ali Sahib Bahadur of the Hindu Mahasabha noted 'look at Article 275 along with Articles 226–227 and 229. The centre can legislate for all the provinces on all matters. And look again at the long list of Union and concurrent lists…all these clearly show that in the hands of a central government which wants to override and convert the federal system into a unitary system, it can be easily done' (CAD 1948: 296).

Other parties existed, indeed the Indian electorate had a bewildering array to choose from, but they were regionalised to specific states, and marginalised by the hegemony of the Congress party system. At the national level, their only hope in altering policy outcomes was by lobbying likeminded Congress members. Just prior to the first mass election of 1952, the Election Commission of India recognised 16 national parties and 60 state opposition parties. After 1952, the number of recognised national parties was drastically reduced to four, since only four parties had succeeded in capturing more than 3 per cent of the vote. Subsequent changes in the criteria upon which parties were designated 'national' parties led to several fluctuations in their number. In 1968, the Electoral Commission evolved a three-point definition of what constituted an all-India opposition party: (i) that the party should have been politically active for five years; (ii) that the party should have returned at least 1 out of 25 contestants to both the centre and one of the states after excluding the number of lost deposits; and (iii) the party should be recognised in three or more states (Butler *et al.* 1984).

Furthermore, the logic of a first past the post system discriminated against the representation of opposition parties through the wide disparities between percentages of votes polled and the number of seats won, and where opposition is not spatially concentrated within specific constituencies. In every general election since 1952 (including 1984), the Congress failed to get an overall majority of the national vote. Yet the divisions within the opposition along ideological, personal and above all regional lines allowed the Congress to capture a majority of seats in the Lok Sabha. In 1952, the Congress contested 479 seats out of 489, received 45 per cent of the popular vote and took 357 or 74 per cent of the seats. In 1967, in a slightly enlarged Lok Sabha of 520 seats, the Congress secured 41 per cent of the popular vote and took 279 or 54 per cent of the seats. The second largest party in 1967 was the Communist Party of India, which, on capturing 3.3 per cent of the vote, secured 16 seats. The largest group of non-Congress MPs occupying a bloc of 91 seats was made up of independents. In 1952, 25 parties participated in the Lok Sabha, while 50 parties participated in the various provincial assemblies, none with any immediate chance of taking control of a state and none seemingly capable of taking control of the centre from the Congress (Butler *et al.* 1984).

Thus the nationalism of the Congress government, and the way in which the Congress had evolved under the British as an all-India movement, marginalised the electoral base available for any subsequent all-Indian opposition, and the secular ideology noted by Jaffrelot (although less hegemonic than dominant) structured political mobilisation in ways that were beneficial to Congress incumbency (Jaffrelot 1996: 106). Despite the implicit belief by the Constituent Assembly in the viability of a two-party or multiparty system after independence, needed to animate the democratic parliament, opposition to the Congress would appear, not on the opposition benches of the Lok Sabha but between the party and the government wings at the centre, between Congress national governments and the state governments, and finally between state Congress committees and their respective state legislatures. Such a system gave to a dominant party an inherent flexibility, which disguised the degree of conflict within the Indian political system, as well as the extent to which the formal separation of powers within the Constitution, between the judiciary, the executive and the legislature, depended on the informality of consensus and cooperation within the Congress system itself. All of these dynamics would change dramatically with the onset of independence.

Once independence became a political reality, Parliament and the prime minister began to assert their power over the party structures. The power of the prime minister over the parliamentary party and the Council of Ministers was emphasised vis-à-vis the party president, and the AICC, through the workings of the prime minister's secretariat. Installed to monitor ministerial performances and route communications, where necessary, through the prime minister, the secretariat acted as a sort of 'grand council of the Republic'. Its coordinating functions between the executive and the ministries were increased by Lal Bahadur Shastri (prime minister 1964–1966) and were further increased, along with those of the cabinet secretariat, under Indira Gandhi. The relationship between the prime minister's secretariat and the cabinet secretariat was considered to be administratively complementary in that the prime minister coordinated policy increasingly sidelining the central committee structures of the AICC. Conflict and disagreements occurred between the organisational and government wings of Congress, not just over policy resolutions by the AICC but also over particular areas of responsibility (Morris-Jones 1957: 171). In 1948 the meeting of the AICC was dominated by complaints from the organisational wing aimed at the parliamentary party. Between 1946 and 1951, and in the wake of the resignation of Tandon, the office of party president was to be held by four individuals. Establishing the supremacy of the prime minister over party president required Nehru to hold both posts for a time until finally passing it over in 1955 to a candidate who would readily accept the weakness of the party presidency (Franda 1962: 248–60).

At the state level, the relationship between the governmental and organisational wings of the Congress were more conflictual than at the national level because the divisions occurred less over ideology than the distribution of patronage. Given the nature of Congress dominance in the states after independence, factions not accommodated within the legislative wing of the state assemblies would consolidate themselves within the Pradesh

Congress (or state) committees under the state party president. From this position, dissenters within these PCCs would actively attempt to undermine the MLAs and even the chief minister through their control of party membership and the nomination of party candidates for elections. These disputes, carried out within the overall structure of the Congress, would also, by the early 1960s, extend through the panchayat raj system down to the block and village levels.

The ability of the state party units to co-opt and foster local patron–client systems intensified competition within state Congress units. This tendency became particularly marked once states became organised along linguistic lines, and once village institutions under the Community Development Programme became incorporated into state politics. Significantly, state PCCs pushed for the linguistic re-division of the states long before the Congress parliamentary party had come around to accepting this policy (Franda 1962: 250). The well-entrenched factionalism at the district and the state level under rival loyalties within the Congress made it imperative for chief ministers to assert themselves over their respective party committees and presidents. Kochanek remarked that 'chief ministers wanted to ensure their control of the state by selecting where possible only loyal candidates, thereby minimising the potential strength of dissident factions within the legislative party' (Kochanek 1968: 292).

This they were able to do by using their greater access to political patronage vis-à-vis the central government to dominate state committees. Patronage was used to construct carefully balanced local and district alliances and represent these within the assembly parties and, in particular, the Council of Ministers. This consolidation of state-based machines thus involved the subjugation of state party units behind established chief ministers. Once this had been carried out, chief ministers naturally dominated the higher forums of the national party because of the way in which delegates were drawn or elected from the state parties to the AICC and from the AICC to the various bodies such as the Congress Working Committee, Parliamentary Board and the Elections Committee. This process of institutional consolidation at various levels of Indian society was a critical aspect of the Congress system, which thus incorporated federalism as an aspect of party structure.

The linguistic re-division of Indian states under the pressure of regional lobbies was to increase the power of the states as significant forums of political action and to enhance them as forums in which dominant caste groups could coalesce to effect policy outcomes. The effect of this reorganisation was to give state politics a more intense regional character and to give the states a significant level of societal power (Ray *et al.* 1982: 13). The level of state autonomy from the centre – bolstered by the structuring of the Constitution – allowed the chief ministers, by working through the organisational wing, to negotiate with New Delhi over specific state affairs, centre-state relations in general, and to run their own state election campaigns. Stanley Kochanek commented in the mid-1960s that 'in the past decade, the Congress Party has seen the development of strong provincial leaders whose co-operation is essential to political stability, economic growth and the continual existence of the Party. The result has been an increasing tendency towards the development of autonomous state parties' (Kochanek 1968: 265).

This was the key to Congress system – a powerful set of chief ministers able to insulate their state from central state intervention (and from each other). For all of Nehru's charisma, his personality disguised the extent to which his power lay on regionalised Congress machines, operated by powerful political brokers (Brass 1965: 2). The centrality of Nehru within the Congress detracted attention away from the necessity of state support and the regionalisation of power within the Congress state parties. Morris-Jones commented at the end of the Nehru era, 'the political system had fallen into firm shape, its leading builder was also its greatest disguise, and with his disappearance its character stands out more clearly' (Morris-Jones 1967: 118).

How did democracy and electoral politics relate to dominant social interests, and through the Congress, society as a whole? How do the empirical observations of the Congress system in action, with reference to specific policies, square with the debates on state and society relations touched on in Chapter 1, and in particular within much of the democratisation literature generally? The electoral importance of locally situated patron–client networks co-opted within the Congress had structured the post-independence relationship between party and government at both the centre and state level. This process of aggregation, in which Congress co-opted local notables who could, through existing patron–client networks, deliver votes to the national Congress (often at odds with their immediate interests) was a critical electoral dynamic of the Congress system as well. Had such interests captured the state? Were they synonymous with state power? Again, the Congress system is best understood through a state–society model in which the state is a myriad of forms contested and occupied by differing interests at differing levels of the political system. Dominant interests are best conceptualised through caste–class linkages, agrarian and industrial, rural and urban, with a large intermediate class situated in the merged and merging rural and urban hinterlands (Harriss-White 2003: 43–71). These social interests were situated within the state at differing levels of the overall political system.

The Nehruvian vision of a modern, industrial India, resting on a planned economy, and state targets touched upon differing societal interests and elites in obviously different ways. Although not hostile to capital and to the logic of private accumulation, the state was committed to intervention through nationalisation and through regulation, especially with regard to land reform and aspects of affirmative action. The national leadership was prepared to coerce social interests as much as it was to collaborate with them when necessary or even, *in extremis*, compensate them. The exact success of each strategy depended on the ability of the state to mobilise social forces at differing levels of the federal system, and the ability of opposing interests to co-opt and work against such policies often within the state institutions themselves.

As such, the reliance of Congress on co-optation of local elites risked policy failure, but to conceptualise this in terms of the state exclusively representing such interests is too simple. It is difficult, in reality, to separate the state from society in a multi-layered polity such as India, with a political elite that was fragmented ideologically as well as being spatially diverse (Sarkar 1989). At the regional level, political and bureaucratic institutions came to rest on a series of dominant social caste groups that mediated, distorted and often ignored the wider reformist agenda of the national state and the agendas pursued by middle class urban interests. In respect to Bardhan, it does seem helpful to identify these as part of a dominant coalition, if not one explicitly identified neatly in class terms. Khilnani refers, by the mid-1990s, to three 'cultural segments' which had by then differentiated out to supporting differing party formations (Khilnani 1997: 193). Yet it does not follow that this coalition of societal interests was as much running the state as simply being constitutive of interests that the central state could neither ignore nor change.

In turn, and to complicate matters further, social power needs to be conceptualised in a broader sense than simple material assets and a struggle to accumulate economic wealth. As noted in Chapter 1, seeking to understand political and economic dominance free of a religious or 'a moral economy' has become increasingly problematic, in part because religious forms of social identity have proliferated, and also because the state has been seen to condone, accommodate or even promote forms of religious legitimacy (van der Veer 1994; Zavos *et al.* 2004). While earlier, modernist approaches to political economy saw religion as a form of irrationality that would decline over time (a view best exemplified by Nehru), more recent scholarship has sought to identify the ways in which India's religious pluralism may give structure to the economy, but also define ways in which the structures of economic accumulation provide the means through which religious identities reproduce over time (Harriss-White 2003: 173; Das *et al.* 1999).

As discussed in Chapter 2, religion persists because the state has, through its constitutional and social practices, encouraged it to do so, but it has also persisted because it informs an important 'spiritual' identity, seeking to reconcile 'accumulative success with spiritual health' as well as mediate and structure forms of social competition for scarce resources over an intensely tessellated caste system. Socio-economic change throughout India has been accompanied by religious change and reform, not determined by any dichotomous base/superstructure crudity, but by the meanings attached to material and political structures by India's religious communities. Given the position of the National Congress elite under Nehru, strengthened by the premature death of Patel, religious anxieties were mediated within elites by the consensus imposed by the definition of secularism by the central state, and largely through the intervention of Nehru himself, but they persisted over time (Brown 2003).

With particular reference to Hinduism, within the regionalised hierarchies of the caste system, there existed a meshing together of material wealth and resources (defined through access to land and cattle), access to political resources, and the ascriptive nature of a given caste group (Bayly 1999: 1–24). Contemporary historiography on caste and jati clusters have, in contrast to earlier orientalist studies that stressed their 'timeless' and 'stateless' existence, identified the social and economic transitions that have changed and transformed the system over time (Quigley 1993). Susan Bayly's work critically examines caste, not as some essentialised Indian or Hindu religious cultural artefact, but as a set of varied and contingent responses to wider changes within the subcontinent's social and political order (Bayly 1999). Far from being a rigid and unchanging system of social hierarchy, the overall system has been extremely dynamic with a degree of upward (and downward) mobility, as accumulation and ritualised status have been recombined in innovative, dynamic and explicitly political ways.

The proximity between status and socio-economic power within jati groupings, and its overall relationship to the political system as a whole, is usefully summarised by Srinivas' notion of a 'dominant caste'. For a caste to be dominant, it should own a sizeable amount of arable land locally, have strength of numbers and occupy a high ascriptive status within the local social hierarchy (Srinivas 1987: 10) It should be able to control and maintain complex vertical caste and jati coalitions through socio-economic patronage and ritualistic status. Dominance by a particular jati in one village may not correspond with the dominant caste of the region or district, and dominance has changed over time as caste associations brought jatis together to improve their ritualistic status as a way to economic and material advance, or as a means, after independence, to lay claim to a disadvantaged economic status to claim affirmative action (Srinivas 1987).

While the state does not structure these particular hierarchies, it provides institutional and procedural forms of governance (pre-independence census takings, post-independence Backward Castes Commissions) that caste and jati can use to improve their economic status and, in opposition to wider secular norms, realign their religious or ascriptive status accordingly. Again, the emphasis here is not on a direct correspondence between social power (dominant interests) and the state, but a blurred, porous, contested boundary as the Indian state structures reach the bottom of the Indian federal and administrative system. In an important study published in the late 1980s, Frankel and Rao make a crucial distinction between dominance and state power:

> The term 'dominance' is used to refer to the exercise of authority in society by groups who achieved politco-economic superiority, and claimed

legitimacy for their commands in terms of superior ritual status, or through alliances with those who controlled status distribution. By contrast, the term 'power' is used to refer to the exertion of secular authority by individuals appointed or elected to officers of the state who claimed legitimacy, under law...within their territorial jurisdiction.

(Frankel and Rao 1989: 2)

A whole series of essays revealed the extent to which dominant social groups – defined by jati and local caste clusters – had been challenged by the state's authority and analysed the differing regional strategies adopted by them to retain their power in the face of state hostility, usually by co-opting political machinery and representative institutions, and offering themselves as the key operators in the consolidation of vote banks. Democracy did not simply accommodate already existing dominant interests; it compelled them to change their strategies of retaining access to political state power, primarily at the local and regional level, and the Congress system became the prime forum in which this struggle took place.

It is possible to illustrate the general dynamics discussed so far by setting out a brief overview of Congress policies towards agriculture after independence (Desai 1986; Mellor 1968, 1976; Wood 1984a). In seeking to bring about widespread changes to landholdings, the national government sought to legislate a social revolution under the auspices of a democratic state through a series of enforced land-ceiling Acts, and the redistribution of land acquired by the state through compulsory purchase. In doing so, it consciously struck at the heart of the Congress system, and the consolidated party bosses that ran Congress state machines, controlling vertical inter-caste vote banks. Prior to 1967, it has been argued that one-sixth of the population delivered the vote of the Indian electorate (Manor 1980). Although central government remained institutionally powerful, and state dominance was contrary to power at the local level, 'the centre [could not] impose its will upon the state constituents unless it [was] reasonably congruent with power relations within the state. The centre [could] persuade, advice, induce and warn, but it [could] never dictate'.

The actual degrees of crystallisation varied enormously across India at any one time. Where there was prior convergence between the institutions of the central state and dominant local interests, Congress policies worked relatively successfully and state governments were relatively progressive. Where it diverged, powerful struggles took place within the political institutions of the local state, often mediated upwards through the courts as litigation ended up before the Supreme Court. Eventually, dissent would, by the mid-1960s, involve voting for non-Congress parties aimed at capturing local state power.

In the first instance, the courts became the site for the contestation of Nehru's national consensus on land reform which lay at the heart of Nehru's concept of social democracy. Land reform was a necessary precondition for the industrial planning experiment of the 1950s and 1960s (Brass 1994; Frankel 2005; Gadgil 1961; Hanson 1966; Rudolph and Rudolph 1987). Yet when the Congress began to implement policy the government found itself frustrated by an interpretation of fundamental rights that frustrated essential 'socio-economic reforms aimed at eliminating appalling injustices'. Zamindari reform bills were either struck down on the grounds that they violated Article 13(ii) and the fundamental right to property, or were complicated by the need to offer compensation for the nationalisation of assets. Expensive compensation occurred even when legislation sought to exempt the government from financial responsibility. In states where Zamindar interests were still relatively well entrenched, land reform effectively stalled. In other areas it stopped when it reached the equilibrium between dominant social interests and the political arithmetic of Congress electoral support (Barrington Moore 1993).

In Uttar Pradesh, Bihar and Madras, compensation (associated by social violence against beneficiaries of the reform) compelled the national Congress leadership to amend the constitution, armed with retrospective action, a strategy condemned as damaging and anti-democratic by the opposition. The First Amendment to the Indian Constitution introduced a device known as the 9th Schedule. Articles placed within the 9th Schedule were shielded from judicial review. Any act placed here could not be struck down on the grounds that it was inconsistent with, or took away, any of the fundamental rights guaranteed by the Constitution. Nehru sought to empower the state to break through conservative interests that had utilised the courts to frustrate social policy, but it implied to many an authoritarian tendency that alarmed sections of the political elite, who saw the government undermining the centrality of property to the fundamental rights. (Brass 1994; Corbridge and Harriss 2000).

Other attempts to break through the disjuncture between dominance and power involved an emphasis on institutional and local democratic initiatives. During the decade of independence the government pushed ahead with the Community Development Programme. This was associated with a Gandhian-inspired village uplift programme, which envisaged the creation of a series of village-based, participatory institutions with some executive powers devolved from the state governments. Through this arose a wider National Extension Service, and eventually a scheme of panchayat raj. Divergences occurred between the centre and an explicit ideological commitment to land reform, and state governments, that were increasingly under the control of dominant landowning caste groups who now fought to control local government initiatives. The central government's ability to intervene directly into agriculture was compromised by it being entered into the state lists of the 7th Schedule. When, in the wake of the 1952 elections, the central government sought to impress upon Congress state parties the need to continue with the commitments to land-ceiling Acts, it discovered that in a majority of cases, state governments were dependent on the very agrarian interests they were seeking to reform. Jannuzi noted in Bihar that 'so deep were the divisions within the ruling Congress Party concerning the proposed legislation that the government of Bihar was unable for several years to develop sufficient support (in the assembly) to pass any version of the law limiting agricultural holdings' (Jannuzi 1974: 162).

It was this very blockage that the panchayat raj system was aimed to break. In 1957, the Balwantray Mehta Report, submitted to the Lok Sabha and the National Development Council, urged the government to provide radical democratic decentralisation within the units of local administration, and to co-opt where necessary local and national politicians to push for agrarian change below the level of state politics. Guidelines were issued from Delhi and the central cabinet, but again left in the hands of state governments to implement. Again the success or failure of this legislation was largely determined by the existing patterns of dominance within the states in question (Frankel and Rao 1989).

By the early 1960s, it was widely recognised that such attempts to deepen democracy and to institutionalise grass roots support had largely failed. Hanson remarked in the mid-1960s that the institutional reforms had degenerated to the point whereby existing social elites had simply excluded the intended beneficiaries through the very regulatory framework set up to help them (Hanson 1966; see Harrison 1960). It was again recognised that without widespread and prior social change within the localities, experiments in radical democratic institutional decentralisation would not deliver the goods. Rather they would assist the further consolidation of affluent elites. Iqbal Narain remarked that the malady affecting pachayat raj lay in the nature of the socio-economic linkages between rural elites, local state support and the inability of the Congress – dependent on state Congress parties for national mandates – to break through vested interests: 'the linkages that the rural elite develops with the district and state elite are a political resource that help build up, and more importantly, help preserve, the elite structure at the village level' (Narain 1976: 231).

As an electoral machine, Congress could not force the issue without the risk of widespread defection of their key operators, violence or mass evasion. The Planning Commission, recognising that the limits of structural reform had been reached, recommended a move towards technical innovation to increase existing output through market incentives (Byres 1998; Harriss 1982b; Mellor 1976). Under the influence of reports emanating from the World Bank and the Ford Foundation, panchayat raj institutions became less a framework for radical democratic reform than as a scheme to undertake marketing outlays and to extend credit and seeds for what would become, through the Intensive Agricultural District Programme, the 'green revolution' (see Kohli 1988: 73). The state had been compelled to change its extractive and collaborative strategies (Frankel 1971; Gough and Sharma 1973).

High Yielding Varieties (HYVs) required substantial state subsidies both in terms of pesticides, irrigation and fertilizers, as well as price support schemes through the Agricultural Prices Commission (APC). Although aimed at ensuring producer returns just under the free market rate, political pressure on state governments often led to higher procurement prices than those recommended by Delhi, and resulting upward pressure on the market rate. The APC itself recognised the cyclical nature of this relationship and argued that in the absence of state ownership in the agricultural sector, it was impossible to achieve the cooperation of dominant agrarian interests at the state level without sanctioning general price increases – a move that would be perceived as against the poor.

The green revolution is, in outline, an illuminating insight into a dynamic aspect of state–society relations. Confronted with an inability to push through structural change, the state's strategy shifted away from attempted land redistribution through popular participation towards a technocratic solution aimed at increasing production for the market. In part this shift in strategy can be seen as compromising earlier commitments to social-ism, despite serious and indeed often strenuous attempts by a national leadership to deliver results. What is revealing – and the example here can be reproduced for a whole series of events such as the state's involvement in industrial and commercial licensing – is that state–society relations are not a crude 'layering' of causality, but a complex constellation of power, varying considerably from place to place. Moreover increased agricultural subsidies certainly increased tensions with industrial interests who saw the agricultural lobbies taking an increasingly large share of scarce finan-cial resources, a cause of tension that would become dominant in the late 1970s.[1]

How were these setbacks perceived outside the dominant social coalitions, and their varied and uneven control of state institutions devised to allevi-ate poverty? Despite these setbacks, the various discourses of redistributive policies and the halting attempts at land reform did act as a slow solvent on dominant interests, and innovations of the local state with reference to new types of institutions did empower groups to contest local notables and the wider structures of social power that cemented local economic, religious and cultural hierarchies together (Kotz *et al.* 1994; Kurian 1986). The creation and mobilisation of social and cultural movements – armed with the whole paraphernalia of societies, clubs, newspapers, journals, supported schools and such like, and influenced by socialism, religious reform societies, and the growing importance of caste associations – meant that concepts like the democracy, socialism, secularism and social pro-test were not locked into serving dominant interests alone (Brass 1997; Hansen 1999; Khilnani 1997; Mitra 1992; Mitra and Chiriyankandath 1992; Somjee 1979).

Even if judged to be a failure at the time, the agrarian tensions and con-flicts generated by the complex interaction of state and society through competitive politics, both within the Congress and the emergent party system as a whole, were to be slowly transformative. By 1970, land declared surplus for distribution was around 2.4 million acres, about 0.3 per cent of the total agricultural area of India. Some redistribution to the poor did take place, and Mellor, citing Ali Khusro, believed that under the first five year plan, land distribution had increased the number of owner occupiers from 40 to 70 per cent of the total, but as an asset land remained remarkably skewed throughout the population (Mellor 1968).

Through the fissures and ruptures that opened up within the socio-economic base of the Congress system, a space emerged that was to facilitate both the rise of non-Congress state parties and, more critically, the opportunity for a central populist leadership. The differential mobilisation of India's various social hierarchies, dalits and adivasis, and then later the OBCs into positions of political dominance, would have the most profound electoral consequences for the Congress and for the capacity of the party and the state to govern. In short, socio-economic, cultural and religious identities began to differentiate, providing separate sectors for more specialised social and regional support. These localising networks required Congress state parties more than ever, to calculate their own local hierarchies of dominance and adjust state power to accommodate these as much as possible, and as much as the centre would allow. Where this failed, opposition parties were quick to insinuate themselves into local structures of power and, where possible, state governments through contesting the Congress. One such party was the BJS.

Although oscillating between the sangathanist strategy favoured by the RSS, and the Congress-like approach to co-optation and election favoured by Mookerjee, the BJS by the early 1960s had adopted a populist strategy related to a concept of integral humanism that sought to combine some appeals to the RSS with a coherent ideology pitched at smallholders, small traders and the petty bourgeoisie (Graham 1990; Jaffrelot 1996). Recent ethnographic studies have noted how socio-economic change generated competition within Hinduism, through the tessellation of caste and sub-caste groups around new and innovative occupations, and between Hinduism and minority faiths. This competition manifests itself not just in communal violence over economic resources between differing faiths (Engineer 1984; Engineer and Shakir 1985) but also over the need to create homogenized ethics over divergent forms of Hindu practices, particularly the rise of the lower caste groups to positions that threatened the spiritual and material status of forward castes (Brass and Vanaik 2002; Hansen 1999; Harriss-White 2003). The BJS began to make electoral inroads where it could seek to dislodge from the Congress system what Graham has called the 'middle world' people who often eked out a precarious existence between a more prosperous middle class and small holders and landless labourers (Graham 1990: 153).

This essentially petty bourgeoisie constituency was open to the BJS, but less in the late 1960s on the basis of direct religious appeals than support for their economic and social precariousness amid wide-ranging social change.

Ethno-religious symbolism was in part deterred by a clear commitment to 'police' secularism at the national level, but also because the specific conditions necessary to use it effectively for mass electoral appeals had not yet arisen. As such, the RSS core of the BJS, confirmed following the resignation of Mauli Chandra Sharma, had continued to articulate a policy of integral humanism, and had played a significant part in the cow-protection movements prominent in several north Indian states where they proved specifically and often locally conducive. Thus despite emphasising some of the corporatist and syndicalist ideas found within *Hindutva,* the electoral successes of the BJS lay more in its populism, its opposition to corruption, and the party's robust nationalism over the war with China. Communal rioting had shown a marked increase in the 1960s, and there is some evidence that Hindu–Muslim violence in northern India was provoked by an improvement in the economic status of Muslims (Jaffrelot 1996: 164). Yet deterred by the vigorous prosecution of rioters through the courts, the BJS stressed socio-economic issues, such as its opposition to wholesale takeovers of the grain trade, planned agricultural cooperatives and the extension of the public sector. In 1967 in particular, it benefited from a close association with princes, especially in western India, who felt that the Congress was becoming 'too socialist' and who, as notables, were still able to determine voter turnout in specific localities.

On economic grounds alone, acting as champions of the emerging 'peasant proprietors who had benefited from the 1950s land reforms' brought some electoral gains for the BJP, but it also intensified competition with regional kisan parties, such as the BKD in Uttar Pradesh, and also the more established communist parties (Migdal 1974; Nossiter 1988). While the RSS tolerated the logic of party competition as such (including princely cooptation, especially where their status also involved local religious patronage to Hinduism), the RSS remained a significant impediment to the BJS in that it was not prepared to concede pragmatic seat adjustments with parties to unseat Congress. Moreover cooperation with the communists would remain at this stage an anathema to RSS activists. While Jaffrelot's argument that secularist ideology remained influential at the centre in policing communal activity is essentially correct, the more compelling evidence is that in terms of defining voting interests and intentions, religion was not as yet a significant factor on its own, or a prime signifier of a more generalised anxiety over identity, wealth and ritualised status.

The deinstitutionalisation debates of the 1980s detailed the collapse of the Congress state-based machines, both as a form of governance (insulating regional politics from national) and as vote-gathering machines able to deliver large majorities for the national Congress Party (Brass 1994; Kohli 1990). The implication of such literature is often that they declined through the personal machinations of prime ministers – particularly Mrs Gandhi. In fact, the dissolution began much earlier, through the logic of democratic government itself. In the abstract sense, and to rephrase Frankel and Rao's observations about the dichotomy between dominance and state power into slightly more anti-foundationalist language, democracy encouraged a split between power and legitimacy not just with reference to 'authority' and the state, but with reference to what constituted the nation, creating a novel space for new emergent forms of ideology. Through the democratic revolution 'a new form of secular power now came into being, derived from "the people" which remained however abstract and unrepresentable' (Hansen 1999: 21). For Hansen, the paradox is that democratisation problematised the notion of how to represent an increasingly divergent social order, while at the same time maximising political strategies that seek to legitimate the use of power for specific – and divergent – goals.

The impact of democracy on local patterns of caste and class-based dominance was, in cruder empirical terms, to corrode vertical forms of alliance and mobilisation and facilitate wider and more horizontal mass movement. In the long term, the logic of democratic politics increased the numerical significance of castes, away from the various forward and intermediate castes towards the scheduled and lower castes. Frequent elections and exposure to political propaganda resulted in a gradual, geographically uneven trend towards political awareness that seemed to prefigure the rise of sectional class-like politics, especially within agrarian societies stratified by uneven access to land and social resources (Migdal 1974).

As horizontal mobilisation increased, older forms of political patronage likewise eroded as wider mass appeals became more practical and more politically expedient. This process was to have profound changes for the nature of the Congress system and the ways in which it worked as a surrogate for multiparty democracy. The 1960s were characterised by growing competition at the state and district level over access to government patronage through the regulatory structures of the state itself, and access to state subsidies. The period was characterised in the north by the gradual consolidation of the intermediate cultivating castes. This brought them into direct conflict with the upper or forward castes who, in places such as Bihar and Uttar Pradesh, remained politically powerful with significant patronage over the lower castes, and well established within the state Congress parties. The resulting conflict led to serious factional instability with Congress state parties and the eventual breakdown in the electoral basis of the Congress vote (Brass 1994; Frankel 2005)

Politically fickle middle castes made it easier for the non-Congress parties to mobilise electorally significant support on the basis of ideology, class and occupational interest (Ahmed 1970: 979). In the early 1960s one observer noted that 'there appears to be a sharp contrast between the types of leadership groups, the traditional village notable and the emerging middle peasantry, which are increasingly found to be vying for control of the local Congress organisations' (Kochanek 1968: 351). Frankel argued that in conditions of ongoing socio-economic change the multi-caste alliances led by traditional landowning patrons and constructed with the support from the lower caste group were becoming more and more difficult to sustain (Frankel 1971: 197). These increasingly articulate (and unmediated) interests were to be further politicised by the trauma of war (with China in 1962 and Pakistan in 1965), and between 1964 and 1966, by drought and concerns over food security that translated into renewed interests in price support, subsidies and land reform.

The dilemma for Congress was illustrated by a series of notable by-election setbacks in previous strongholds and, after 1962, an erosion of the Congress vote generally (Butler *et al.* 1984) translating into gains by the opposition, especially the BJS. There was a palpable sense of concern within the party, as early as 1963, that both conservative and socialist parties would outflank the Congress. In 1962, the BJS made significant gains in New Delhi and within the Hindi belt. In response, there followed several attempts to make Congress state units more representative through weakening the hold of the bosses by bringing party units into a closer organic relationship not merely with Congress legislatures, but with the organisational interests – student and peasant associations already at work within society (Morris-Jones 1967: 109–32).

This concern about representation within the Congress at a time of enhanced electoral competition also explains the Kamaraj Plan, a suggestion by the party president, Kamaraj, that important congressmen from the government wing should resign their positions and go to work in the organisation (Zaidi 1983: 377). While it was clearly aimed at renewing commitment to the party organisation (independent of the spoils of government office), it was also aimed at weakening the grip of powerful politicians, rooting out corruption and eliminating chief ministers that Nehru considered hostile to his ideological vision.

Following Nehru's death in 1964, the sense of political disarray intensified. Congress chief ministers were keen to preserve their autonomy and press on with a specific, regionalised approach to politics, arguing that they were best able to recognise the intricate and increasingly complex socio-economic alliances needed to gain votes. After Nehru it is clear that the dominant prognosis within Congress, was to strengthen the component parts of the old system – the party machines run by the chief ministers – and thus favour a 'weak' prime minister who would leave in place the de facto regionalisation of the national party.

Potential candidates for the post included: the Home Minister and acting interim Prime Minister, G. L.Nanda, a well-known Hindu traditionalist; Morarji Desai, also a Hindu traditionalist and a well-known Gandhian; Lal Bahadur Shastri; and Indira Gandhi, who was then broadcasting and information minister. Once Shastri had been decided upon as fulfilling the required specifications, the Congress president went about mobilising support mainly through the AICC and the CWC and in consultation with individual chief ministers. The parliamentary party was largely sidelined. Morarji Desai (who was known to be ambitious), when informed of the consensus achieved over Shastri's candidacy within the party as a whole, withdrew from the contest and Shastri was elected unanimously. Michael Brecher remarked at the time that 'the key role of the chief ministers in the selection of Shastri gave them greater leverage with the centre. Closely related was the lesser stature of the prime minister' (Brecher 1966: 65, 1969)

However, the sudden death of Shastri in 1966 at Tashkent in Soviet Asia presented a renewed crisis of leadership at a time of economic and political crisis brought about by the second Indo-Pakistan war and a growing slow-down of the Indian economy. Unlike 1964, there was no chosen candidate, but Kamaraj was able to canvas various chief ministers and extracted from 8 out of 14, an early indication of support for Indira Gandhi, defined in the main by their continuing desire to keep Morarji Desai out of office. Moreover, as a junior minister she was believed to be a relative novice, a powerful symbolic continuity of her father without his determination or skill (Kamal 1982: 99). Morarji Desai insisted on contesting the leadership after refusing to accept a CWC compromise. A vote took place in the Congress parliamentary party with Indira Gandhi receiving 355 votes to Desai's 169. Desai's supporters were both regionalised and ideologically biased to Congress traditionalists.

In the cases of Indira Gandhi and Lal Bahadur Shastri, the Congress parliamentary party was 'guided' in its choice over the succession, despite the parliamentary party's implied right via the republican constitution to select its own leader and thus, de facto, appoint the prime minister. Once more the party had nominated a prime minister who was seen as being devoid of political ambition, identifiable through her family name while promising to be politically malleable. Her first cabinet was constrained by party considerations, the implications being that the Congress Party president expected to be consulted on matters that had conventionally become, after the mid-1950s, the priorities of the prime minister. There were, however, early indications that Mrs Gandhi had a slightly different role in mind for the office of the prime minister.

How significant was Mrs Gandhi's emergence as prime minister? 'Reading' political elites into wider social formations and clusters of 'power', and relating these to institutions of governance is now (rightly) perceived to be a contingent business, more about deduction than prediction. Calibrating into this equation an emphasis on political leaders themselves becomes a difficult but necessary part of understanding the contingent nature of political action. Individual leaders matter, even if they do not (and should not) provide the starting point for the analysis of an entire political system, and at particular moments of crisis, they can matter enormously. Analysts have noted the importance of Nehru outliving Patel, and the significance this had for India's peculiar views on secularism, even if he never managed to escape the legacy that Patel left within the party. Much has been written about Indira Gandhi, and much stress has been placed on the idiosyncrasies of her childhood, and the difficulties she experienced in her early adult life.[2] While Mrs Gandhi is not responsible for the ongoing differentiation that continued throughout the Indian polity, she undertook a series of political decisions that structured the actual form this differentiation would take, and in part the consequences that would follow.

In 1966, Mrs Gandhi had showed her ideological inclinations towards secularism in her response to the Cow Protection Movement, much in keeping with her late father's attitude. Contained within the directive principles, the Congress was committed to eventually abolishing the practice on the grounds of animal husbandry, not the principled grounds that it offended the sensitivities of Hindus. Ritualised cow slaughter, which formed a rare Muslim ceremony marking the end of Id, was a controversial issue that had often, along with loud musical processions by Hindus outside mosques, provided the catalyst for communal riots in areas where Muslims were perceived as significant local competition (Brass and Vanaik 2002). Mrs Gandhi dismissed G. Nanda from his position of home minister following a speech in which he identified with the position of the BJS in some of the northern states, especially UP (Jaffrelot 1996). The prime minister presided over an Emergency Committee that sought to prevent sahdus from using Gandhian fasts to gain publicity and pressurise the government, and again raised the prospect – touched on by her father but abandoned shortly before his death – of legislating to ban communal parties and the political use of religion. In her speeches she picked up and reiterated the themes often used by her father regarding the sinister nature of the RSS and its shadowy links with the BJS. Other acts of political unilateralism quickly followed. In 1966–1967, while chairing a cabinet meeting to discuss the current economic crisis, the prime minister secured a vote to devalue the rupee by just over 57 per cent without discussing the matter with the party, and having also overruled the concerns of the finance minister, Morarji Desai. The decision created a sensation, along both ideological and procedural grounds.

In 1967, Mrs Gandhi presided over the worst national performance of the Congress since independence. Although still able to form a comfortable majority, opposition parties, particularly the BJS, made significant gains. Taking just under 41 per cent of the national vote, a decline of around 4 per cent from the 1962 elections, Congress returned to power with 283 seats, losing 78 MPs to a welter of differing parties (Brass 1965: 72). The BJS took a record share of the vote (9.35 per cent) and a record number of 35 MPs. In the state elections its vote also improved, and it secured 268 MLAs from a total of around 3,487 contesting across the Union, emerging as the principle opposition party in Uttar Pradesh (Jaffrelot 1996: 213). In immediate contrast to the Nehru years, the instability and chaos that resulted in the wake of the 1967 election created concerns and anxieties about the future stability of orderly, democratic government in India. Even where Congress' electoral support held up, the workings of a first past the post electoral system combined with serious factionalism within the Congress to produce unstable and opportunist anti-Congress governments. Faced with coalitions held together merely by the promise and largess of office, some political analysts and reformers responded to the ensuing scramble for office by calling for reform to the electoral system and even questioned the suitability of mass democracy for solving India's social and economic problems.

The immediate political fallout in the states was instructive. Seriously weakened chief ministers attempted to run fractious administrations by creating new posts and dividing others. When such coalitions broke down, as they often did, Congress legislative parties proved as capable as non-Congress parties in inducing defections, particularly, if the Congress appeared to be on the brink of being returned to office. Congress willingness to take in 'untested and often unknown' individuals for opportunistic reasons was as widely condemned as it was practiced. For the first time governors were confronted with competing claims to office by political leaders, all able to prove a majority on the floor of the assembly. In such cases it was not uncommon for the rival lists to contain the names of the same MLAs, a testimony either to their pragmatism or their ambition.

Some rather notorious cases are instructive. In Uttar Pradesh, for example, of the two lists of supporters submitted to the governor by the Congress and the SVD coalition after the 1967 polls, 13 names were on both (Siwach 1979). Frequent re-elections after 1967 led to continual defections of MLAs across party lines, both before and after the polls had closed and there were lengthy periods of time before any government emerged. On 25 June 1968 in Bihar, the governor called upon the Congress leader to form the government despite the much-publicised fact that 84 MLAs listed as supporting the government had changed their party affiliations at least twice (Ghosh 1970: 66). The Policy Planning and Research Division of the Union Home Ministry reported that 438 registered defections took place in the year following 1967, compared to the 542 defections registered during the previous decade (Kamal 1982: 127). In such circumstances, administrations did not last long, and governors (and through them, the offices of the prime minister and the Union president) were continuously involved in making (and often as not, unmaking) governments.

The most immediate consequence of this sudden exposure of the governor was a sudden rash of President's Rule. During the 11 years of Mrs. Gandhi's first two terms in office (1966–1977), President's Rule was declared on 27 occasions, compared with the 7 occasions during Nehru's 14 years in power. In a majority of cases between 1967 and 1969, the excuses for central intervention were legitimate – the collapse of the constitutional machinery at the state level – but in almost every case some aspect of the operation was suspect. In situations where the incumbent government was non-Congress, the pressure on Delhi to act came from Congress Pradesh committees and Congress ex-Chief Ministers, although often as not, in circumstances of constant flux, the Governors' instinct was to return Congress to power where ever possible (Siwach 1979). Such innate conservatism, more so than evidence of a central conspiracy, led to opposition outrage.

For example, in Rajasthan in 1967, a 13-party coalition presented its claims to the Governor to form a government. Mohan Lal Sukhadia, the former Congress Chief Minister, also staked a claim at the head of a coalition. After some deliberation the Governor called Sukhadia to form the government on the grounds that, although he had less support on the floor of the assembly than the United Front, the United Front was relying on independents (Siwach 1979: 289). This was a curious precedent to set, and the Governor's decision was condemned throughout India. As Surendra Mohanty was to state in 1973, 'it is not for the Governor to foreclose the issue or to act under the surmise that, if an alternative government were formed, it would soon fall' (LSD 1973: 228), however reasonable such a surmise appears with hindsight.

More seriously, political instability within the state governments exposed the general ambivalence within the Constitution over the exact relationship between the governor and the chief minister. Was the governor always to act on the advice of the chief minister (Kurian and Varughese 1981: 11)? If not, under what circumstances could this advice be ignored? Was the governor obliged to allow chief ministers an opportunity to test their majorities on the floor of the assembly before dismissing them? The opposition were quick to point out situations in which the chief ministers were not given time to prove their majorities in the assemblies before being dismissed. In some cases they were dismissed merely on the grounds that the Congress leader could list more followers – often from the ranks of the government they were seeking to overthrow. The unfolding dramas between the governors and the chief minsters were watched with particular fascination in part because of the general expectation that they would set precedents for the relationship between the president and the prime minister once coalitions manifested themselves at the centre, as many expected them to do.

Such disputes over procedure – dangerous in any formal system of representative democracy for bringing into disrepute wider issues of governance and legitimacy – led to novel constitutional problems. In West Bengal, the speaker of the assembly retaliated against the 'unfair' dismissal of the previous chief minister by continually adjourning the assembly. Riots and disorder within or outside various legislative assemblies became commonplace. Opposition parties petitioned the president of the Union to replace specific governors who had continued to act in the interests of the Congress Party. Forced to allow an emergency debate on the role of the governors in UP and Bihar specifically within the national Parliament, the government came under renewed pressure to reform the system or to issue specific guidelines. Without these improvements, it was charged that 'the governors are gradually being reduced to instruments of the ruling party' (LSD 1970: 305), echoing a concern that had been expressed within the constituent assembly.

The other consequences of political instability – predominantly throughout northern India – were that large areas of political decision-making became the responsibility of the bureaucracy (*Economic and Political Weekly*, 18 March 1967: 545). State bureaucrats found themselves directing state policy in the various interregnums of popular government. In some cases, the resumption of bureaucratic responsibility restored administrative stability, but generally the extent of political instability severely damaged policy implementation and the scope for policy initiatives, as well as obviously restricting popular access to decision-makers. The frequent collapse of governments in Bihar and Uttar Pradesh – particularly Bihar – led to a marked deterioration in law and order. In a curious testimony to the commitment of the higher bureaucracy to the supremacy of elected politicians, few senior civil servants were prepared to make initiatives during the various interregnums of popular government. Moreover, frequent re-elections were expensive and often violent, with a notable increase in incidents of ballot stuffing and attempted vote rigging.

At the centre, the consequences of the elections were indirect but serious, and it soon became evident that they had serious consequences for Mrs Gandhi. Between February 1967 and February 1968, 14 state governments fell, 6 of them having been under the previous control of the Congress (*Christian Science Monitor*, 3 April 1967). With the myth of the electoral invincibility of Congress having been exploded, many of the non-communist parties began seeking national coalitions or mergers with like-minded parties in anticipation of national elections and pre-electoral seat adjustments (*The Times of India*, 15 January 1969). In the meantime, the number of non-Congress state governments within the federal system raised the risk of there being open confrontation, within the federal system, with a Congress centre (*The Free Press Journal*, 3 March 1967). All of this was uncharted territory, as was the prospect of a coalition government at the centre after 1972, and a non-Congress prime minister working with a Congress president (*Hindustan Times*, 1 May 1967)

So intense was the resulting anxiety over instability that, in 1967, a meeting of the CWC went so far as to consider the question of coalition governments and the 'threat of military takeovers about which there has been some loose talk in the country' (CWC, 7 May 1967). Significantly, the rupturing of the Congress system reversed the priorities of the Congress chief ministers overnight. On 16 June 1964, in a regular column of *The Statesman*, Inder Malhotra noted that prior to the elections, the Congress chief ministers throughout the party wanted a weak prime minister to effectively keep out of regional affairs. After their defeat in the states, they wanted a strong prime minister committed to removing non-Congress parties proving to be inept at restoring the status quo ante. In order to make the Congress fit enough to resume government, the prime minister had to show decisiveness. It was their singular misfortune that Indira Gandhi, a Prime Minister selected for her weakness by the syndicate – the collective term popularly used to describe the chief ministers within the party structure – was now condemned for failing to take the initiative.

It was to be against the background of the fourth national elections, and the conflicting explanations for Congress's performance, that the battle between Indira Gandhi and the syndicate was to be joined. An enormous amount has been written on this struggle, and the resulting Congress split (see Chatterjee and Mallik 1997; Hardgrave 1970: 256–62). The salient points of the crisis are reiterated here merely to reveal the significant change that it brought about in the way the political elite, through the state, sought to act on society and manipulate or change the patterns of consensus and conflict unleashed by the fourth general election. The strategy of socialism, of mobilising the poor, was in the context of a specific threat to the leadership, brought hastily to the fore of governance. It offered an ideological reading of how Congress had lost the 1967 elections, and it also offered a way of saving the Prime Minister.

The AICC's official report on the election reiterated the view that there had been no clear ideological intentions behind the result and thus no general failing of the Congress Party (INC Report on The Fourth General Election 1967). The setback was put down to general incumbency and a need to 're-invigorate' the party structures, especially the role played by party members. Yet, significantly, there were sections within the party that argued that Congress had been castigated for its failure to deliver previous socialist commitments. A small group of powerful politicians argued that Congress decline was 'terminal' unless there was a clear reorientation of policies to take account of what appeared to be in reality a more sceptical and restless electorate, more demanding and less easily manipulated. Concern was also expressed about the rise in the BJS vote in particular, which was linked to the increase in communal violence (INC Report on the Fourth General Election 1967: 66).

The Congress Forum for Socialist Action (CFSA), formed in 1957 under S. N. Mishra as a ginger group to push for structural economic reform, stated categorically that the Congress must assume an uncompromising left wing identity. Associated with the CFSA's early activities were Chandrasekhar and Krishna Kant, both young members of the Congress parliamentary party. The CFSA successfully campaigned for a new 10-point programme at the June 1967 meeting of the AICC in New Delhi (AICC Resolution 19 June 1967), calling for outright nationalisation of the top 14 banks, the abolition of state pensions to ex-Princes (the so-called Princely Purses), and a new wave of agrarian reform and ceiling legislation to prove that the Congress stood with the rural poor and was prepared to deliver them tangible gains. Many senior members of the party, above all Morarji Desai, resisted the call to radicalise government policy. The syndicate stood in growing opposition to what it saw as a group of disaffected former socialists and communists working within the consensual traditions of the Congress (Frankel 2005).

Initially the prime minister stood apart from the ongoing ideological disputes within the Congress Party. Although the prime minister and some senior cabinet ministers received the 10-point programme favourably, it did little to impress the cabinet with a sense of urgency. Such apparent diffidence gave rise to the concerns within the syndicate that either the prime minister was sympathetic to the extremism of the Young Turks, or was unable to discipline them. There was much talk and speculation over the future of Mrs Gandhi. In February 1967, *The Hindu* stated that 'most of the confusion on India's political today is directly attributed to the uncertainty surrounding the choice of the next Prime Minister' (*The Hindu*, 1 February 1967).

The debate on the scale and direction of the economic policies of Congress continued inconclusively until the 1969 midterm polls in West Bengal, Uttar Pradesh and Bihar and Kerala, brought about by the need for fresh elections. The indecisive results in Bihar and Uttar Pradesh further indicated the extent to which the Congress state parties' electoral credibility had collapsed into factionalism. Appeals to national unity by the centre and threats of disintegration could not by themselves ensure the electoral support needed to restore stability to the Congress. This continual failure of the Congress at the state level cast shadows over its prospects of retaining control of the centre after 1972. On 1 March 1969, *The National Herald* argued 'there is no Congress leader who can declare with confidence that the Congress Party will be able to retain the position [at the centre] after 1972'.

Poor showings at the polls intensified the policy debate within the Congress, with the Young Turks renewing their call for radical, socialist policies. At the Hyderabad Session of the AICC in 1968, the syndicate retaliated by effectively stage-managing the organisational elections to seven seats, which was interpreted as move to isolate the prime minister. (*Hindustan Times*, 5 June 1968). Added to this, the newly appointed Congress Party president, Chief Minister of Mysore [Karnataka] Nijalingappa, was a firm syndicate man and closely identified with the right wing of party. It was at Hyderbad that the first intangible rumours began that there was a plot to remove Mrs Gandhi. These rumours persisted until they came into sharp focus following the sudden death of the Union President, Dr Zakir Hussain.

The vacancy of the Union presidency altered the opportunities available to Nijalingappa to constrain or even remove the prime minister through presidential intrigue by nominating a Congress candidate loyal to the syndicate, who, given the likely prospect of a hung parliament after the next election, would be able to act decisively. The resulting selection of Sanjiva Reddy as Congress candidate for the Union presidency was widely interpreted as the beginnings of the syndicates' long-awaited coup against the prime minister. As a staunch member of the syndicate, and associated with Desai, he was known to be openly hostile to the Young Turks and their calls for nationalisation, and critical of Mrs Gandhi's ineffectual government. Mrs Gandhi, travelling throughout East Asia, was reportedly 'informed by two cabinet members that the party bosses had decided to press for the nomination of Sanjiva Reddy with the ultimate aim of toppling the government' (Krishnan 1971: 303). On her return to Delhi, she thrashed out a strategy to openly challenge the party's choice.

In an interview with his biographer, Y.B. Chavan remarked that, from the onset, the Prime Minister was convinced that this conspiracy was real. Frank has provided further evidence of this interpretation (or obsession) in a more recent biography (Frank 2001). The Prime Minister opposed the nomination of Sanjiva Reddy as Congress presidential candidate, attempting initially to enlarge the franchise responsible for his selection, and finally suggested a compromise candidate. Confronted by the prospects of facing Reddy as President, Indira Gandhi aligned herself with what had been from the onset a separate ideological debate within the party and the country as a whole. In effect, though her 'infamous stray thoughts' paper in 1969, she co-opted the radical economic programme of the Young Turks and converted into an ideological platform on which to confront the syndicate. She did so, conscious of the fact that her power struggle would be fought through the idiom of socialism versus conservative reaction (Krishnan 1971). After the Congress split in 1969, Piloo Mody (Congress-O, the Congress organisation under Nijalingappa) argued during a no-confidence motion against the prime minister and the council of ministers that 'the entire economic programme of the government has been based upon some stray thoughts, rather hurriedly dictated, which miraculously transformed themselves from the gluttony of the Young Turks into an economic programme which is supposed to herald in the great socialist dawn. Why?' (LSD 1970: 305)

Compelled by the fear of a conspiracy against her, Indira Gandhi turned to already existing analysis, arguing that the electoral weakness of the Congress was tied substantially to the socio-economic failings of state leaders and the unrepresentative character of state machines that had ignored the poor. The Prime Minister also accepted the belief that successful economic reform required a centralist government actively participating in state affairs because these had been compromised by vested interests. Moreover only if the Congress became electorally more viable and the bosses removed would she be able to secure her own position as prime minister at the centre (Gangal 1972: 39).

Aware that Nehru had been locked out of various policy areas of his own government, and if not policy, then excluded from implementation, Mrs Gandhi argued that chief ministers were no longer capable of ensuring electoral turnouts or political stability in the states and as such they were dispensable. The opportunity existed within the turmoil of the post-1967 situation to reinstitutionalise the Congress system excluding the chief ministers and their machines as independent sources of patronage. This would require placing political power in New Delhi around the Congress

Parliamentary Party and the cabinet, and depriving the state parties of critical independence. It would involve seriously curtailing the power of the organisation as a separate series of institutions. Curtailing their patronage would weaken their bargaining vis-à-vis New Delhi and open the way for direct mediation between New Delhi and the Indian electorate which now appeared impatient for change.

Having staked out her ideological position, Mrs Gandhi needed a candidate that would appeal to dissenting Congress legislators and attract enough non-Congress votes within the presidential electoral college to seriously threaten the chances of Reddy winning. Again, the sheer number of non-Congress MLAs within the college made the syndicates' job of managing a successful win on the first ballot difficult. It was both a crisis and an opportunity. The decision by V. V. Giri – the vice-president and, following Hussain's death, acting president – to stand as an independent gave the prime minister the opportunity to support discreetly a rival candidate (Sathyamurthy 1969: 480). The fact that Giri was in some way a radical with a proven Congress background and closely connected with the trade unions gave further credence to any 'ideological' interpretations of the prime minister's move against Reddy and the 'reactionism' of the syndicate (see Krishnan 1971: 309). More importantly – if not somewhat procedurally dubious – as acting president, Giri was in a position to use presidential decrees to assist the prime minister and, indirectly, his own chances of victory.

Mrs Gandhi deployed a political *blitzkrieg* against the syndicate that seemed to many out of character with the cautious, diffident prime minister of 1967. Moments before he resigned as acting President and stood for election, Giri issued an ordinance nationalising the top 14 banks and insurance houses of India. Mrs Gandhi then removed Morarji Desai from the Finance Ministry on the grounds that she felt him 'ideologically' opposed to her policies. He subsequently resigned as Deputy Prime Minister. Although constitutionally legal, the issuing of an ordinance was seen by her opponents as being underhand, especially given the fact that it was issued two days before the Lok Sabha reconvened for its Monsoon session. As a normal bill, nationalisation would had stalled in protracted debate, and even if passed (which seemed extremely unlikely) it would not have allowed Indira Gandhi to assert dramatically an ideological issue through one swift, bold executive act. The Prime Minister defended the move on the grounds that advanced notification of nationalisation would have encouraged financial manipulations aimed at undermining the ordinances' effectiveness, yet in one move, the embattled Prime Minister had extended her campaign into the streets as a champion of the poor.

The Supreme Court strengthened the identification of Indira Gandhi with a radical socialist policy by subsequently striking down the ordinance on the grounds that it was *ultra vires* of the Constitution, a move she depicted as reactionary and one carried out at the instigation of the syndicate

(Austin 1999). To make matters worse, Nijalingappa appealed to the legislatures from the Swantantra and the BJS in a search to gain votes for Reddy, a move that was quickly interpreted as the embracing of communal elements within Indian society, and which Mrs Gandhi was quick to condemn.

Mrs Gandhi then pursued an interventionist strategy in the states to gain the support of Congress MLAs for Giri's candidature, and, implicitly, for the continuation of her own radical prime ministership. The states nearest to New Delhi, such as Uttar Pradesh, Haryana, Himachal Pradesh and to an extent Punjab and Rajasthan, were particularly vulnerable (*Hindustan Times,* 28 July 1969). Uttar Pradesh was to receive special attention because of its strategic position in the Hindu heartlands of northern India and because of the large number of votes allotted to it in the presidential college. Indira Gandhi held a number of talks with Charan Singh in Uttar Pradesh – a former congressman with a powerful Jat farming base in the state – aimed at persuading him to return to the party. Throughout these discussions, the Congress chief minister – C. B. Gupta, a staunch supporter of the syndicate – made several formal protests to the Congress Party president that such discussions between his prime minister and one of his key rivals was aimed at nothing less than toppling his government.

In July, Nijalingappa demanded that the Prime Minister issue a whip to Congress MLAs, instructing them to vote for Reddy (Quraishi 1973: 126). Indira Gandhi avoided this by calling for a 'conscious vote' – since what was at issue was the future direction of the party. She argued that the nomination of Reddy raised moral questions pertinent to all congressmen and they must have the liberty to search their own hearts and minds before coming to a decision. She also argued that the issuing of a formal whip would adversely influence the non-Congress candidates' chances, something she claimed would be undemocratic (Zaidi 1968–1969: 694).

This policy enabled the prime minister ample scope to encourage Congress cross-voting without contravening party rules. Chief ministers unsympathetic to the prime minister could not but help favour calls for a free vote because they could not enforce a whip without a grave risk of mass indiscipline. Consequently V. V. Giri became President mainly through non-Congress votes and dissident Congress votes in heavily fractionalised states. On the first count Giri secured 48.01 per cent with Reddy on 37.49 per cent. The third candidate, C. D. Deshmukh, secured a mere 13.48 per cent and was eliminated from the second ballot. This was the first time in an Indian presidential election that second preferences were counted to secure that the winning candidate crossed the threshold of 50 per cent of total votes. The ideological orientation to the conflict was confirmed in startling detail when Reddy took the support of syndicate congressmen and MLAs from the Swantantra and BJS, while Giri took Congress votes loyal to the prime minister, plus a variety of socialist and CPI votes on the second preferences. Although Charan Singh had been influential among the BJS and the Swatantra leaders in persuading a third candidate to stand, the BKD's vote in Uttar Pradesh in the second ballot was vital for seeing Giri through. Revealingly, following the announcement of the result, Indira Gandhi defended her action on the grounds that she was restoring the powers of the prime minister (*The Times of India*, 22 August 1969).

A formal party split now became unavoidable (Chatterjee and Mallik 1997; Hardgrave 1970: 256–62). Even before the ballot, 30 MPs had a private meeting with Nijalingappa and complained that, given her refusal to issue a whip, the prime minister must be disciplined by the party and resign from the prime ministership (*The Times of India*, 2 November 1969). At the same time, however, 60 AICC members signed a petition demanding the resignation of the party president. Chief ministers from seriously divided administrations were the only real constituency with a vested interest in working out a compromise because their administrations would fall if Indira Gandhi and Nijalingappa parted ways in New Delhi.

There were some unity efforts, headed up by Y. B. Chavan, but they could not detract from the infighting going on throughout northern India for the control of state parties, fought out in the full glare of the media. On 23 August 1969, Indira Gandi had made a call to all 'like-minded people' who had left the party to return to the fold. This call was a tactical one based on the assumption that after the nationalisation issue had been resolved, ex-socialists could now be talked into rejoining the party. It was also based upon the calculation that opportunistic elements with significant followings at state level – people like Charan Singh in Uttar Pradesh, Rao Birenda Singh of Haryana and Govind Narain Singh of Madhya Pradesh – could be persuaded to bring strategic factions over to the prime minister's assistance. In early October, in retaliation for the expulsion of two of the prime minister's staunch supporters from the CWC (C. Subramaniam and F. A. Ahmed on the grounds of 'unpardonable deceit'), a signature campaign was launched to reconvene the AICC and to remove Nijalingappa from office. Mrs Gandhi set up a rival CWC at the prime minister's residence at Safdarjang Road.

After the Congress split at the centre, the prime minister's faction (the Congress-R) was left with about 40 per cent of the organisational strength of the party, and 441 out of the 750 district committees. This division did not occur neatly within dissident states. Some district committees rebelled even though the state units as a whole remained with the Congress-O. The all-India divide between the two central factions resulted in a geographical concentration of the Congress-O in the south and west. The majority of the Congress parliamentary party rallied about the prime minister with 222 of 283 sitting MPs 'defecting' with her and 114 of the 154 Rajya Sabha MPs coming over to her (Hiro 1977).

The split in the numerous state congress parties (and governments) had a dramatic impact on political stability. Chief ministers loyal to Mrs Gandhi sought to reconvene party organisations after they were expelled, and vice-versa. Chief ministers who remained loyal to the syndicate found that a parallel Congress-R party organisation sprang up that sought to enlist MLAs and expel others. Presidents and vice-presidents of Congress Pradesh committees suspended and dismissed each other from the Congress, while the prime minister sought to use central patronage as an incentive to encourage waivers to stay loyal or to rediscover their loyalty. Once in the open, rival factions took various disputes to the courts over property, ownership of membership lists, even the party emblem. The central offices of the Congress Party in Delhi were subject to lengthy litigation that was only resolved in the early 1970s.

It is difficult, with hindsight, to exaggerate the impact of the split and the consequences it had, not just for Congress, but also for the functioning of the wider constitution and the national institutions of the state, which suddenly found themselves disconnected from the traditions of the Congress system. For the first time in Independent India, a political drama was played out before a mass political audience, coded in a straightforward narrative of radical socialist versus reactionary conservative. As later became clear, in the context of an already eroded Congress system, this realignment was to prove critical for the fortunes of the Congress-R and compel the opposition parties, including the BJS, to regroup against Mrs Gandhi either on the dubious grounds of defending privilege or over communalism, or the much more problematic issue of seeking to outbid the Congress on a populist ticket aimed at the poor.

The immediate concern of the prime minister after December 1969 was to shore up what had become a minority government at the centre. By 16 November Indira Gandhi had obtained the support of the CPI and the DMK on the floor of the Lok Sabha. On the basis of the 10-point programme and the nationalisation of banks, the CPI considered the Congress-R to represent the radical or 'progressive wing of the bourgeoisie' and extended support on the understanding that the Congress would not reinstate any preventive detention (following the lapse of an earlier statute) and that the Congress would push for the immediate implementation of the economic 10-point programme and work closely with the CPI throughout north India. With the central government temporarily secured, Indira Gandhi was keen to re-establish stability within states already under her control, or destabilise states that were under the control of the Congress-O, such as Gujarat and Mysore.

To do this, she encouraged unprecedented levels of political defection aimed at bringing down unsympathetic governments. In most cases, this technique was assisted by the use of President's Rule to allow Congress assembly parties time to ensure sufficient support to form the administration. In some cases, President's Rule, for which in many cases no governor's report was ever submitted, was used to blatantly remove non-Congress governments. Again Uttar Pradesh provides a clear example of how constitutional procedures could be manipulated to open up opportunities for state Congress parties (*Economic and Political Weekly*, 26 September 1970: 1584). Uttar Pradesh and the infighting between Mrs Gandhi and Charan Singh for control of the state government revealed how immensely useful a complicit Union president was for a determined prime minister running a coalition government at the centre.

As the immediate elite nature of the split receded, there was the urgent need for the newly reconstituted Congress to re-engage with its social base and to build up its support in the run-up to elections in 1972. Having dismembered the old party, Mrs Gandhi realised that she headed what was, in many crucial respects, a faction without a coherent institutional or organisational base. The working committee of the Congress-R, meeting in Delhi on 20 December 1969, had discussed the urgency of reconstituting itself throughout India and the need to draw up an economic policy that took as its departure the 10-point programme, first passed at Bangalore. Where land reforms had been discussed and land ceilings agreed, chief ministers had been requested by the interim party president of the Requisitional Congress (C. Subramaniam) to report back on implementation by 31 March 1970. The Congress president also convened a new Parliamentary Board.

The Board decided that organisational elections would be held as soon as possible and that in the meantime Pradesh Congress committees badly damaged by the split would be dissolved and ad hoc committees formed under central supervision. On 11 December 1970, the CWC decided to postpone organisational election for five weeks. At a subsequent meeting at New Delhi on 19 December, it was decided that 'in view of the large numbers of Congressman returning to the Congress fold and the ongoing court case concerning the Party symbol, it has been decided to postpone the organisational elections for an indefinite period (Zaidi 1970–1971: 257).

The Congress-R's AICC meeting in New Delhi on 13–14 June 1970 pursued the discussion concerning economic policy. In the light of the unprecedented activity by the Swantantra and the BJS in league with the Congress-O, the AICC emphasised the need to be on the guard against 'right wing, reactionary' parties seeking to undermine the government's economic policy. The old guard had shamelessly sought the support of communal parties that 'were against the best traditions of the party'. Mrs Gandhi was able to deftly produce evidence of a creeping 'conspiracy' aimed at preventing her revolutionary policies from being implemented to reinforce the general view that the Congress-O and the anti-Indira coalitions were 'right wing', a term that implied both communalism, and the conservative views of the Hindi traditionalists that she saw personified by Morarji Desai. The open association of the Congress-O and the BJS, prefiguring the later relationships of the mid-1970s, seemed to some analysts to imply an emergent two-party system premised on ideology, not opportunism.

Earlier that year, on 11 April, the Supreme Court upheld the quashing the presidential ordinance nationalising the banks, to the evident jubilation of the Congress-O leaders. Later, on 10 October, the government bill preparing the way for the abolition of princely purses – the last item on the 10-point programme – was defeated in the Rajya Sabha by three BKD votes and then struck down by the Supreme Court following an attempt to pass it by presidential ordinance. On 30 September 1970 the prime minister had failed to convince a meeting of chief ministers in Delhi to lower the ceilings on land legislation and speed up implementation. This was the first serious indication that, as the membership of many PCCs had weathered the split, many Congress state governments were still capable of excluding the centre from areas deemed to be their specific concern or go against their own political interests. Significantly, in light of the mounting problems over implementing the party's economic policies, the CFSA decided not to dissolve itself as a lobby within the Congress-R. The CFSA decided to continue to push for a radicalisation of politics on the grounds that, once the political necessity of socialist support had passed, the prime minister might be tempted to compromise on various socialist policies (Torri 1975: 1077–96).

The weakness of a coalition government, the hostility of views in the upper house and above all the power of the courts to use property rights as a basis to undermine government economic policy highlighted the weakness of ruling by ordinance. The only way to break the deadlock within key institutions of the state was for Mrs Ghandi to regain a parliamentary majority in her own right. In late December 1970 in a televised broadcast to the nation, the Indian Prime Minister announced that the Lok Sabha was to be dissolved and that the Congress-R would seek a fresh mandate for implementing its socialist policies (*The Times of India*, 28 December 1970). After a detailed study of the Congress' performances in several Lok Sabha by-elections and the Rajya Sabha elections in 1970, the Congress Parliamentary Board had decided that substantial benefits could be achieved at the state level through a central victory. This and the decision to bypass finally the various intermediate elites of the Congress led for the first time to the formal splitting of central and state elections.

This decision to de-link the Lok Sabha elections from elections to the state assemblies was to transform state–society interactions dramatically. It was initially based upon the assumption that, once the political battle had been resolved at the centre, state parties would come round to the winning side through the logic of central political patronage (interview with D. K. Barooah 1985). In many senses, this reversed the logic of the old Congress system that Morris-Jones had commented on in the late 1950s. State elections were to be dominated by national leaders and national themes. This was particularly true in the case of harijans, the minorities and specifically poor agrarian interests throughout the Indian subcontinent whose votes could be cobbled directly into a national Congress mandate. An article in *Hindustan Times* entitled 'Emergence of lobbies' argued that scheduled castes and tribes constituted about 60 per cent of the Indian population and controlled about 182 Lok Sabha seats. A strategy that aimed to capture these could be used to restore the Congress to power. Moreover separate national elections would allow the Congress to emphasise further its ideological appeal in states where left-of-centre parties had already proved successful in mobilising low caste votes (*The Times of India*, 17 August 1967)

The call for elections caught the Congress-O opposition largely off guard. Their efforts to form alliances or permanent coalitions had continued after the 1969 midterm polls, encouraged by the apparent certainty of a coalition government at the centre by 1972 (*The National Herald*, 28 April 1969.). Apart from approaching the Congress, the BKD had turned to the BJS and the Swatantra in an attempt to forge a pre-poll coalition, a risky strategy given the strident ideological language that was being used in Delhi to justify India's first midterm national election, and given the fact that in socio-economic terms, the BJS and the BKD were competing for the same votes. Attempts by the non-communist parties to merge were undermined by their attempts to retain their specific regional identities, and by tensions between the BJS and the RSS over the extent to which electoral appeals needed to address socio-economic, as opposed to ethno-religious issues. RSS cadres baulked at the prospects of joining state governments containing communists, and the retention of youth movements and other organisations from merger with similar movements within the opposition led to renewed anxieties over 'entrism'.

The campaign of the resulting 'Grand Alliance' betrayed its opportunistic strategy by focusing entirely on the personality of Indira Gandhi and the dangers of 'communist infiltration into the government'. It was countered by the Congress-R's eminently successful slogan of *garibi hatao*. The campaign was nothing short of an ideological coup. At a CWC-R in New Delhi on 6 January 1971, the manifesto was drawn up, noting in its preface that 'the 1967 elections had registered the people's impatience with the pace of development throughout India'. It then drew attention to the directive principles of the Constitution that committed the 'State to strive to promote the welfare of the people through adequate distribution of economic goods' and committed itself to doing this within the confines of respecting and maintaining the institution of private property (General Secretaries' Report 1971: 516). The manifesto went onto state that 'it will be our endeavour to seek such further constitutional remedies and amendments as are necessary to overcome the impediments in the path of social justice' (General Secretaries' Report 1971: 516).

The policy outlines of the manifesto consisted of a commitment to small and marginal landholders/cultivators, security of ownership and an extension of credit linked up with the earlier nationalisation of the leading banks (Torri 1975: 1077–96). Such electoral appeals were targeted at lower caste voters, especially at scheduled castes and tribes. While they did not explicitly exclude the agrarian middle castes, the CWC-R appreciated that in order to marginalise the middle caste support of various opposition parties throughout northern India, they needed to mobilise the lower castes en masse.

Yet they needed to do this in circumstances where middle class urban interests would not be unduly alarmed. While industrial policy included a passing reference to an expansion in the role of the public sector and the imaginative steps behind the setting up of a Monopolies Commission, it reiterated a commitment to the private sector and stated its intentions to simplify regulations placed on large companies to allow them to compete.

In the fifth general election, the Congress-R contested 441 seats and won 352 on 43.8 per cent of the popular vote with the Congress-O winning a mere 16 of 238 seats contested (Butler *et al.* 1984: 1133–52). In effect the Congress-O was reduced to a Gujarat-based regional party. As had been intended, the Lok Sabha result had a dramatic effect upon those states and districts that had retained Congress-O loyalties, initiating another ongoing process of defection, as well as on membership within the Rajya Sabha (Heginbotham 1971: 1133–52). Thus the process of factional realignment continued, further delaying the holding of organisational elections and resulting in the further dismantling and reconvening of PCCs. In April 1971, as the level of instability in the states worsened, the AICC had gone through the lip service of cancelling organisational elections again while calling for the pressing need for 'effective and powerful machinery to be set up at the government as well as the organisational level throughout the country to implement the policies and various measures relating to socialism (Zaidi 1970–1971: 226).

However, the massive mandate of 1971 was reaffirmed a year later when, in the aftermath of the Bangladesh war, state elections were held in 16 states and 2 Union Territories. The Congress Election Committee played a vital role in awarding party tickets, breaking the previous dominance of the Pradesh committees over candidate selection. In view of the continuing alliance between the CPI and the Congress-R, the central Election Committee sanctioned the Pradesh committees to make some electoral adjustments at the state level with the communists, particularly in Bihar, Uttar Pradesh and West Bengal (*The Statesman Weekly*, 2 February 1972). The party tickets were awarded to a relatively large number of defectors where they deemed to be of local significance. Implying that the Congress-R was reverting to the politics of aggregation and not cadre-based mobilisation, this move was criticised by the CFSA.

4   The state and political crisis
      1971–1975

In the specific context of Independent India, the Congress split of 1969
brought about a critical change in the nature of state–society relations.
Mrs Gandhi effectively collapsed the party organisation into the government. Changes brought about by the split concerning the relationship
between the president and the prime minister – continued later by
Mrs Gandhi, arguably under the influence of P. N. Dhar and Haksar –
further collapsed government institutions into the state. There was thus a
concentration of power that had all the hallmarks of a new regime seeking
to change the nature of the state and to alter the ways in which it sought
to collaborate with and extract resources from dominant interests situated
within society. This perception, widely held by the opposition after 1971,
and at times deliberately and manipulatively used by the government, created
an atmosphere of crisis, which many identified with the precursor to some
form of fascism, or from the perspective of the BJS and the Grand Alliance,
some form of communist coup. Yet such a strategy – to be exemplified by
the Emergency itself – was incoherent, reactive to wider events and at times,
almost random.

The collapse of the party was to transform the entire nature of the
Congress, from the recruitment of MLAs and chief ministers, to the selection of prospective Lok Sabha candidates, and to the ways in which party
funds were collected and disbursed and campaigns organised. Whatever its
weaknesses as an independent institution (and in comparison with the ideologically based and cadre-based organisations like the RSS and some of the
Marxist parties, these had been many), the elimination of the party proved
to be catastrophic. This was largely recognised by the government itself, as
indicated by the sheer number of times it was discussed within the newly convened AICC and brought up in separate communications between the prime
minister and the chief ministers with reference to the implementation of land
ceilings and agrarian reform – communications that revealed how confused
Mrs Gandhi was as to where the party ended, and where the government now
began (LSD 1972: 212–15; see General Secretaries' Report 1973).

As the 1972 party elections were to reveal, ideological and institutional
coherence had been sacrificed to the urgent, pragmatic consideration of

letting former repentant Congress workers and politicians back into the fold where their support was needed (*The Times of India*, 23 December 1972). As a result, one of the great ironies of the split was that it destroyed organisational coherence without actually changing the nature of the elites at the state and local levels: in doing so it created the first outline of the disjuncture between the institutions of governance and those they claimed to empower. *The Hindu* commented cynically that the organisational elections returned old-fashioned, pre-split Congressmen to power (*The Hindu*, 26 December 1972). It certainly failed to resolve the issues that the split rhetorically claimed to be have been about.

Restructuring the party increasingly relied on a form of ad hocism, in which temporary and short-term expedients quickly replaced the need to bring about a much more complex process of institution-building (*The Hindu*, 22 February 1973). Despite retaining a separation between the prime ministership and the office of the party president, Mrs Gandhi ensured that all future heads of the party were subservient to the government. In 1969 she appointed Jagjivan Ram, a staunch ally of hers during the crisis days of the split. The prime minister ensured that any attempt to renew the party organisation was done under her supervision and in such a way as to prevent any threats to her continuation in office. Mrs Gandhi was to emerge as a decisive leader, but only in a crisis, and one of her own making (interview with Haksar 1985). Into this vacuum fell a series of senior Congressmen, and a group of powerful chief ministers who were in effect the exact opposite of the syndicate: men who were centrally appointed and at the beck and call of the prime minister, politicians who in many cases did not know their states, and often as not men who were incompetent. Also sucked into the institutional gap that was now opening up between what the government claimed do and what it could, in effect, deliver was the Youth Congress, a curious vehicle for Sanjay Gandhi and his 'over-educated and often out of work' friends that came slowly into prominence during the early 1970s (Kothari 1983a).

This over-personalised network of linkages between Congress workers and the government, and through a huge majority and a complacent Union president, powerful aspects of the Indian Constitution provided the context for rampant corruption and the use of public assets to make private gains. It was carried out through the abuse of industrial licensing and the regulatory structures of the state, but also through the need to fund what were from now on national election campaigns requiring huge resources to retain power. Ad hocism provided an ideal cover through which local notables and elites could insinuate themselves back into the local state structures. Prior to the onset of the JP Movement itself, several incidents are illustrative of the miasma of corruption that began to cloud up the workings of the central institutions of state – namely the prime minister, the prime minister's secretariat and the cabinet secretariat. In 1971, following a series of bizarre accidents, it became widely known that the government had access to some sort of slush fund held 'irregularly' if not illegally within the Reserve Bank of India (LSD 1972: 271; Jagmohan Reddy Commission of Enquiry 1978).

Moreover, charges of electoral malpractice were levelled at the prime minister concerning her campaign in the Rae Bareilly constituency, Uttar Pradesh, also in 1971. The charges covered 52 separate offences under the 1952 Representation of the Peoples Act. Yet perhaps the most pernicious allegations of corruption were those that gathered around Bansi Lal, a friend of Sanjay, implicated in the infamous Maruti car scandal (Government of India Report (Maruti Affair) 1979), and L. N. Mishra, the union minister for railways. These were to become symptomatic of the wider crisis the government was to face.

Two separate scandals linked L. N. Mishra to the prime minister in New Delhi and to the general political instability of the Congress Party in Bihar. Mishra's alleged criminal activities in Bihar were initially documented by an Estimates Committees Report, and later by the Kapur and Chagla Commissions of Enquiry. It had been alleged that, in his role as finance minister for the Bihar government, Mishra had misused the money raised for an industrial project for party funds and other 'private purposes'. The issue was raised in the Lok Sabha in late August 1973. Mishra, in a personal explanation before the House, dismissed the accusations as baseless and false, drawing attention to the 'political motives' behind the non-Congress state government's attempts to discredit a central minister.

The corruption charges at the centre involved the alleged use by Mishra of industrial import licences to exact 'party donations' from companies. On 8 December 1974 the Central Bureau of Investigation (CBI) issued a report, which in the eyes of the opposition established Mishra's guilt (LSD 1974: 244). The prime minister refused to lay the report before the Lok Sabha on the grounds that it would prejudice ongoing legal proceedings. This led to further accusations of a cover-up. Morarji Desai pressurised the prime minister to allow the opposition leaders access to the report. By late 1974 L. N. Mishra's name had became synonymous with corruption. In May of that year, *The Motherland* (a paper politically associated with the BJS) stated that 'Mishra has become a sink of corruption capable, it is rumoured, of overpowering the prime minister' (*The Motherland*, 5 December 1974). In late December 1974 it was widely believed that Mishra was under increasing pressure to resign (*Hindustan Times*, 21 December 1974; Matthew Commission of Enquiry 1975).

Despite a series of high-level inquiries into these and other incidents, corruption fuelled a climate of secrecy and encouraged the opposition to believe, not unreasonably, in the existence of conspiracies and cover-ups. As early as 1970, the opposition moved the first of many no-confidence motions against the government on the specific charges of encouraging defections in opposition-held states, an excessive concentration of power in the hands of the prime minister through the prime minister's secretariat and the cabinet secretariat, and a general dilution of ministerial responsibility resulting in an overall bureaucratisation of central government. Mrs Gandhi's subsequent majority after 1971 made such parliamentary censure largely decorative, but continued pressure from such a small 'embittered and out of touch opposition' added to her contempt for the processes of the Lok Sabha (LSD 1970: 543). On 16 May 1973 Shyamnanda Mishra (BJS) introduced a motion against V. C. Shukla, Gokhale and Subramaniam – all cabinet ministers – implicating them in the general corruption charges against Sanjay Gandhi and the dealings of Maruti wherein the prime minister had been given access to government funds and soft loans (Report on the Commission of Enquiry 1977: 54–67).

Explanations for the reconfiguration and centralisation of governmental institutions in the wake of the split and almost continuously in the run-up to the Emergency are varied and complex. Rather like the various attempts at party building from 1970 until 1972, it was in part the product of improvisation, of attempts to streamline government and to create a powerful centralised executive that could oversee important aspects of social and economic policy. Mrs Gandhi presided over the dramatic increase in the prime minister's secretariat and the gradual stripping over of departments from the home ministry to create what became, in effect, a new locus of power within the government (*The Indian Express*, 25 September 1969; see *The Times of India*, 9 January 1969; *The Daily Telegraph*, 7 July 1970).

On 9 January 1969 it was announced that the prime minister's secretariat was being expanded to increase its coordination powers over various ministries. The move involved a limited transfer of departments from the home ministry and was defended on the grounds that following the rise of non-Congress state ministries; the Congress Party alone was no longer capable of organising centre–state relations. The prime minister's institutional power was also extended by the retention of various cabinet portfolios while undertaking numerous reshuffles. In 1969, Indira Gandhi controlled the relatively new department of atomic energy as well as the ministries of planning and finance. In 1970 she had added the important home affairs ministry to her responsibilities, as a temporary charge (*Hindustan Times*, 8 July 1970; *The Search Light*, 28 January 1971).

In 1971 the Prime Minister held three cabinet posts, relinquishing finance in exchange for information and broadcasting. Rarely were portfolios reshuffled without having had some of their departments or functions retained by the Prime Minister's secretariat (Kohli 1983). This was especially true of the home affairs ministry. The cabinet secretariat also underwent enlargement as various departments – electronics, science and development – were added, along with the Central Bureau of Investigation and the intelligence bureau. These latter moves led to widespread allegations that the prime minister was spying on the opposition and on members of her own government. Such techniques involved the misuse of civil servants for 'party' purposes, and the use of junior ministers to 'double check' their senior, cabinet colleagues (Dhar 2000; Haksar interview 1985).

This process of 'double clearing' relied on ministers of state who were personally known to the Prime Minister (or increasingly by 1974, to Sanjay Gandhi) and who reported directly to the Prime Minister herself. These were people who were known as 'fixers' – people who would get things done, and who regardless of their formal position within government, had the ear of Mrs Gandhi. Significantly, they reproduced (and strategically reinforced) the improvised, personalised networks that came to stand in for the party organisation at large in the states – a tangible manifestation of the collapse of the party into the government. On a whole series of important issues – from preventive detention to land-ceiling Acts and industrial licensing – relevant ministries were often not informed of their briefs or allowed to access only information that their civil servants deemed necessary. P. N. Haksar and P. N. Dhar as private secretaries to the prime minister had more influence than any member of the political affairs committee on the cabinet, although in the end Haksar fell out over the growing influence and role of Sanjay. This emphasis on loyalty also proved important for appointing chief ministers to run Congress states. On the whole, in a ministerial system oddly reminiscent of National Socialist Germany, Mrs Gandhi encouraged competition and rivalry among her peers and also sought to isolate or discipline ministers by striking at their supporters in their states. Ideas over policies were often sanctioned on the basis that they expressed the desires of the prime minister, and may well have started off as leaked 'intentions' through the prime minister's office (interview with P. N. Haksar 1985; Haksar 1979; Dhar 2000).

The degree of improvisation by the prime minister should not, however, distract attention away from the ideological motives behind her attempts to restructure the Congress, and through it a state able to implement meaningful social reform. P. N. Haksar and Dhar, as well as the CPI, and through them, members of the CFSA, favoured a strong centre because they believed it provided the framework to break through the various areas of institutional resistance at the state level. Mrs Gandhi made it known that she was determined to bring about 'fundamental' changes to the pattern and principle of government to demonstrate that she stood with the poor. Explicit within the 1971 election campaign was a call for constitutional change in order to remove some of the obstacles that had frustrated the government in its attempts to carry out radical policies, especially with regard to the judiciary, whose interventions had particularly irked her. In stressing the importance of the directive principles in the radicalising discourse of the 1969 split, Mrs Gandhi (or rather, her colleagues in the CFSA) rightly identified the difficulties of having invested property within the fundamental rights, and then having prioritised these over the directive principles. The difficulties stemmed from having identified these difficulties in isolation from the need for broader, proactive strategies that required institutional renewal within the party and the government.

Her government's strategies to contain the courts became almost obsessive, especially in the wake of the *Golak Nath vs. State of Punjab*, which had ruled that 'parliament's power to amend the constitution could not be used to abridge the fundamental rights' (Austin 1999: 197). One strategy involved passing legislation to eliminate the court's ability to interpret the validity of constitutional amendment bills, including the fundamental rights, where they involved socio-economic reform. In effect this would amount to undermining judicial review of the constitution by claiming a doctrine of parliamentary sovereignty. Such a strategy was politically dangerous since it risked an all out war with the judiciary, which could, at the onset, strike down such legislation as *ultra vires* of the constitution. Yet it was a high-profile risk that the Prime Minister was prepared to take. The 24th Amendment sought to extend Parliament's right to amend the constitution, including the fundamental rights, and also further restricted any room for the Union president to withhold his assent from any future amendments (Austin 1999: 244). Other acts followed. The 25th Amendment bill, introduced in November 1971, aimed at preventing the sort of legal complication that had beset the nationalisation of the banks and the abolition of princely purses as well as previous land-reform policies. The bill did this by introducing Article 31(C). This Article excluded from judicial review legislation aimed at alleviating economic inequalities.

Another strategy, and in the short term more appealing, was to direct and encourage the court to be socially responsible and to recognise the structural causes of widespread inequality; in effect, to call for a 'committed' judiciary, ready and willing to side with a popular government with a radical mandate (Rudolph and Rudolph 1987). An element of this second approach was also in effect to pack the court when senior appointments deemed to be ideologically suitable (or personally loyal) and to intimidate and threaten the high courts if they proved unwilling to comply with government wishes. The most dramatic example of the lengths Mrs Gandhi could go to set aside convention occurred in April 1973, when the post of Chief Justice of India fell vacant, in the wake of a controversial court ruling that sought to establish a temporary truce with the government by establishing a 'basic structure' doctrine. The *Kesavananda Bharati* verdict, handed down in April 1973, conceded that the fundamental rights could be amended but that the basic structure of the Constitution could not be. The Supreme Court had not gone onto define what, in effect, constituted the basic structure. Yet even this caveat was not enough for Mrs Gandhi. In recommending to the Union president the appointment of Justice Ray, Mrs Gandhi superseded the three most senior justices, all of whom had voted with the majority verdict on the basic structure, and appointed a man who had submitted the minority view supporting the government's case for parliament's right to unqualified constitutional amendment.[1] The government brushed aside opposition outrage by stating that it had been a necessary step in ensuring 'that the socio-economic reconstruction of our country is not going to be interrupted or upset through judicial pronouncements' (LSD 1973: 350).

In sum, and as shall be briefly outlined with reference to the JP Movement, the greatest failure of Mrs Gandhi 'idea' of India was the inability to establish a meaningful link between the institutions of government, government policy and the people who had, through the *garibi hatao* mandate, empathised with and voted for change. The commonest explanation for this failure highlights the opportunism of the government and Mrs Gandhi's lack of interest in pushing for the abolition of poverty once she had secured power. Mohan Dharia, a Congress MP from Pune, had warned as early as 1970 that, if the prime minister failed in her commitment to the Congress (and more specifically the CFSA), pressure would be brought about to remove her from office (*The Free Press Journal*, 18 August 1970). Only in the highly unusual circumstances of the Emergency, when the prime minister was again fighting for her job, did she return to the radical language heard during the split.

Other commentators – principally Francine Frankel – argue that Mrs Gandhi's attempts to re-forge the state into a powerful instrument of social policy by linking it directly to those who would benefit from it was effectively resisted again at the key interface of local and state politics (Frankel 2005). With hindsight, it is difficult to separate out the opportunism of the government from positions of obvious principle – such as Congress' commitment to the wholesale takeover of wheat and rice, and the determination of the government to push forward this policy in the very worst circumstances imaginable. The essence of Mrs Gandhi's populism was a government whose strength lay symbolically with the poor, but which lacked the ability to transform and extract significant concessions from a series of elites situated within society and still able to co-opt and manipulate local state institutions. Jaffrelot cites Shil's definition of populism as the essence of Mrs Gandhi's approach to politics in that it 'proclaims that the will of the people is supreme over every other standard... it exists wherever there is an ideology of popular resentment against the order imposed on society by a long-established, differentiated ruling class, which is believed to have a monopoly of power, property, breeding and culture' (Jaffrelot 1996: 235).

To some extent of course, Mrs Gandhi was a faction of that coalition, whose political survival became, in the context of the 1967 election results, an ideological shift to the Left that conspired to use masses against structures. In a war against her former allies and the wider social interests that sustained them, whatever Mrs Gandhi's intentions were, the consequences of her policies played into the ongoing socio-economic transformation that had been eroding the old patron–client structures of Indian politics since the late 1950s. Yet she did not replace them. In the vacuum created by the party-now-government, over-centralised and reliant on a series of personalised links between the prime minister's office and the states, this new system was precarious and unstable. In retrospect, its only hope of survival lay in its ability to oversee sufficient piecemeal and incremental change to buy off calls for radical change, or to sufficiently empower newer interests to renew local institutions themselves by facilitating a broad social movement. Unfortunately, in the circumstances of a severe economic crisis, stimulated in the main through drought, war and external economic shock, the government was compelled to confront its weaknesses immediately and head-on.

As the economic crisis deepened, the government resorted to a style of operating that had in many senses been finessed during the 1969 split. Maintaining a close relationship with the Union president, Mrs Gandhi retained the habit of issuing presidential ordinances just after the Lok Sabha went into recess, or just before a parliamentary session was to begin. Given the numerical strength of the Congress after the 1971 elections, it was never in doubt that such ordinances would become law when placed before parliament, yet the use of ordinances enabled the government to avoid confrontation with the opposition at the introductory stage of legislation, a stalling tactic that the opposition had perfected since 1971 and which caused considerable ill will. In part caused by their dubious use of parliamentary procedure, it was facilitated in the main by poor quality drafts and general incompetence by the law ministry. When the MISA Amendment Bill was introduced by the home minister on 7 May 1975, the non-communist opposition had the bill withdrawn on a technicality that required a redrafting. Jyotirmoy Basu said revealingly 'I want an assurance that they [the government] will not show contempt for the House by promulgating an ordinance during the intercession period' (LSD 1975: 331). On 29 June the MISA Amendment Act was promulgated and presented to the Monsoon Session of the Parliament in August for a statutory vote. In 1973, 4 ordinances were issued, in 1974, 14 and in 1975, no less than 25 ordinances were issued (Raj 1982: 391–427).

Alongside the use of ordinance, and complementary to it, was the government's increasing reliance on 'special powers' legislation, legitimately devised in one context but rapidly deployed into another. The CPI had, as a condition for joining the Congress in coalition in 1969–1970, asked for the government to scrap plans to introduce preventive detention legislation. After the landslide election, and stimulated by the Defence of India Rules passed in the wake of the Indo-Pakistan war of 1971, preventive detention (MISA) was introduced and retained through a whole series of amendments until superseded by the declaration of the Internal Emergency in June 1975. The government's justification for retaining (and increasing the scope of) MISA, as well as anti-smuggling and hoarding legislation after the end of the Bangladesh war, rested on its effectiveness in combating the economic crisis throughout the country, especially in the context of food scarcity and the high prices for essential goods (Home Ministry Report 1973–1974; see AICC meeting 1974; Zaidi 1983). By 1974 MISA was used effectively against the railway strike to force employees back to work. The *Eastern Economist* commented perceptively that 'the danger in this situation is that the public may easily let themselves be converted to the ruinous assumption that the authority of the ordinary laws and the normal jurisdiction of the law courts should be progressively displaced by what are essentially the executive powers of the government' (*Eastern Economist*, 11 January 1974: 35).

In the wake of both the *garibi hatao* campaign and Indira Gandhi's successful prosecution of the 1971 Indo-Pakistan war, the BJS in particular found itself hampered by a party electoral strategy premised on princes and notables associated with conservative values, and an RSS link that caused mutual suspicion amongst the non-communist opposition generally. An attempted shift to embrace populism in its 1971 manifesto was buried in part by its participation in the Grand Alliance, and in part the size of Mrs Gandhi's majority. Ironically, the BJS had moved towards a populist strategy at the same time as the Congress-R, and was simply incapable of outbidding it (Jaffrelot 1996). The result of this setback was to reinforce the RSS tendency to return to strategies of social networks and to protest against corruption and the immorality of Congress rule through social protest.

Yet the risk of legal repression against the RSS remained a very real one in the wake of Mrs Gandhi's mandate for change. From 1970 onwards there had also been a renewal in the centre's willingness to police communalism, partly in the wake of the agitations over cow slaughter in the late 1960s, which Mrs Gandhi overwhelmingly blamed on the RSS. From 1968 onwards, Congress members close to the prime minister (such as Jagjivan Ram) worked alongside other parties in calling for the banning of the RSS (Jaffrelot 1996: 239). Following allegations that the RSS had been prominent in a series of communal riots – mostly notably Ahmedabad – Mrs Gandhi consulted Y. B. Chavan, her then home minister, over the feasibility of outlawing the RSS completely. Chavan's recommendation was that it would be better to tighten up on the implementation of existing law (with reference to the wearing of uniforms and the public use of weapons in ceremonies) rather than allow the RSS to mobilise sympathy around overt prosecution. This was a tacit recognition that as a cultural movement it was too big to take on directly. Yet in 1972, Indira Gandhi introduced the Criminal Law (Amendment) bill that increased the central government's ability to monitor and outlaw regional paramilitary organisations and to prevent them participating in demonstrations and political rallies.

Yet while the renewed threat of state action, combined with the success of Congress led by Indira Gandhi (Congress-I) in outbidding BJS populism, appeared to confront the Hindu nationalists with a setback, the speed at which the *garibi hatao* mandate collapsed presented it with a unique opportunity. The socio-economic and political conditions of the early 1970s allowed the BJS-RSS to combine the tactics and organisational strategies that had, until now, always appeared to divide it: to mobilise in society around themes close to the RSS, and to combine with other non-communist parties in denouncing the corruption and scandals of the Congress. What Jaffrelot rightly and evocatively calls 'the impossible assimilation' came about because the BJS had identified with interests that had initially supported the Congress-I, but which had become quickly disillusioned – and outraged – with Congress failure. Assimilation with mainstream politics was assured by the launching of the JP Movement and the involvement of Hindu traditionalists with Gandhian-style protests against a morally decadent government. In 1974, having recently replaced Golwalker as head of the RSS, Deoras condemned corruption and called for a cleansing of the nation through selflessness and patriotism. This was a language that was now widely being used against Congress-I, and was especially present within the Congress-O. The RSS believed that the stark failure of the government to satisfy popular demands would assist their own widespread penetration of society, especially if they were able to emphasise the government's immorality and selfishness.

The JP Movement was in the main an attack upon government corruption, but in conditions of famine and massive official incompetence, it took on a moral force that enabled the opposition to stage a confrontation with Mrs Gandhi and with the idea of the Indian state itself. The response of the state was to rely even more on coercive strategies linked to the language of socialism, and to stress even more so the indispensable nature of the prime minister herself. In part the causes of India's economic crisis of the early 1970s lay in the very model of development that Nehru had adopted at independence: an initial over-commitment to industry, and then a later commitment to pro-market strategies within agriculture without any sustained structural change in the size and distribution of landholdings (Harriss 1982a).

In industry, the principle of licensing, of state-led allocations of resources to private companies, and statutory regulations on monopolies and foreign ownership created an environment conducive to rent-seeking, inefficiency and corruption (Joshi and Little 1994). Despite Mrs Gandhi's nationalisation bill of the early 1970s, her government proved unwilling to alter the basic principle in which resources committed to the provision of public goods also (and disproportionately) encouraged the growth of income in private hands (Corbridge and Harriss 2000). A failure to mobilise income through direct taxation – especially from the newly emergent agricultural capitalists – led to a decline of investments aimed at the public sector (Toye 1981).

A poor monsoon in 1972–1973 led to a dramatic fall in agricultural production and compounded scarcities created by the increase in international petroleum prices, and the earlier Indo-Pakistan war. In 1972, agricultural production fell by 9.7 per cent against the previous year's output (*Eastern Economist*, 13 September 1974: 492). Resulting grain scarcity led to a large scale hoarding of wheat across northern India and a dramatic rise in grain prices. By August 1974, wholesale prices were 2.4 per cent higher than in July 1974 and 25.7 per cent higher than the rate recorded in August 1973.

Between August 1973 and August 1974, cereal prices increased by 40.4 per cent, while the price of pulses increased by 43.3 per cent. The annual rate of inflation jumped from 9.9 per cent in 1973–1974 to 22.6 per cent in 1974–1975. By September 1974, the prices for all essential commodities were generally 32 per cent higher than for the same period of the previous year. Economically the only bright spot in 1974 concerned the continuing rise in Indian exports, which since 1972–1973 had been increasing at an annual rate of 22 per cent. The balance of trade was still generally adverse however because of the increase in food imports and the increase in petroleum prices (*Eastern Economist*, 1 November 1974: 817). Despite these conditions, the central Congress remained committed to the establishment of a public distribution system to ensure 'fair prices' for essential commodities, particularly wheat and rice, sugar, edible oils, kerosene and standard cotton yarn that had emerged in debates immediately in the wake of the 1969 split. At the Congress AICC meeting in 1970 it had been argued by leading members of the CFSA that the only way to ensure a public distribution system for cereals was to nationalise the wholesale market in wheat and rice (*Northern India Patrika*, 12 June 1970).

The instability within Congress-run states from 1970 until 1972 postponed any further attempts by the centre to move directly against agrarian capitalists producing surpluses for the market. On 13 January 1973, the government directed the Congress-run states to take over the wholesale wheat trade in the current 1972–1973 season. This was later postponed following disagreements among Congress chief ministers about the level of government price support to farmers, who started a series of agitations to ensure the prices were adequately set for their interests. On 26 February 1973, Fakhruddin Ali Ahmed (the then agriculture minister) made a statement in the Lok Sabha announcing that the takeover would start at the beginning of the next rabi (winter) season. On that date the government would set up a system of single-state food zones and prevent any private trading between grain surplus and deficit states. All purchases in the market would be made through the State Finance Committees to a national 'pool'. Ahmed remarked: 'We have taken a major decision in the national interest for the transformation of the economy involving the interests of millions' (LSD 1973: 251–52).

Such a policy was not popular within many Congress-run states because conditions of drought had strengthened farm lobbies in surplus-producing states, which argued that the procurement prices offered by the government were too low. The BJS – which had opposed earlier attempts to introduce a wholesale takeover of trade in the late 1950s, and which had consistently sought to represent the small traders and the peasant farmers – was presented with an ideal opportunity to gain ground in many of the northern states where it had lost out. In May 1973 the prime minister accused the opposition parties of 'ganging up' to frustrate the government's food policies (*The Times of India*, 15 May 1973), despite the fact that a majority of the states were Congress-run. The AICC pointed out in July 1974 that policies to stabilise price levels 'cannot be successfully implemented without the wholehearted co-operation of the state governments' (AICC Meeting Zaidi 1974: 509). G. K. Reddy of *The Hindu* commented that the 'failure of the wheat take-over is attributed to the indifference and incapacity of state governments to be able to comprehend the complexity of the matter' (*The Hindu*, 22 June 1973), or more revealingly, to directly oppose a powerful set of dominant interests through state power.

By early 1974 it was clear that the government could not procure sufficient supplies of wheat through the fair-price shops to alleviate shortages. This was despite the continual use of preventive detention and special powers granted to the police to threaten and break up hoarding and smuggling. A continual rise in grain prices throughout north India threatened to endanger the cheap supply of grains for the urban consumers and undermine the ability of the state to maintain a regulatory structure. Out of an estimated output of 25.5 million tonnes in wheat, the government only managed to procure 5 million tonnes, of which 3.3 million tonnes came from the Punjab alone (*Economic and Political Weekly* 15 March 1975: 462). By late March 1975 the government had resorted to large-scale rationing in the face of serious food shortages and widespread rioting, and a growing sense of crisis throughout middle class, metropolitan Indian. Confronted with such difficulties, the government withdrew from extensive interventions in the markets. Under a later policy, regulations against wholesale traders were eased, allowing them to make private purchases on the condition that they sold half to the government (LSD 1974: 330). Albeit in difficult circumstances, this constituted a dramatic policy failure.

On 8 February 1974, *Hindustan Times* remarked that the 'spatial dispersal and increasing frequency of disorder and violence have tended to blur public perceptions of the fact that the country is passing through an extraordinary economic and social crisis' (*Hindustan Times*, 8 February 1974). By then, the government had been forced to pass an emergency interim budget and was confronting a series of strikes within the public sector, most notably the railway strike aimed, in the government's view, at paralysing the state. Deflationary policies were aiming to retrench labour and significantly reduce public spending. This palpable and growing sense of crisis was to produce the context in which the government would confront a political crisis, starting in two states that had, in the wake of the split, been exposed to the greatest amount of central intervention.

Central interventions in Gujarati politics were given particular emphasis in the context of the split, since it was one of the few states to retain a Congress-O administration. In late 1970, with active central encouragement, the state unit of the Congress-R began to undermine the majority of the Hitendra Desai-led Congress-O government. In early June 1971, following the Congress-R landslide in the elections, 50 Congress-O leaders defected, including some former ministers, leaders of district committees and some presidents of the local Panchayat Raj system (*Economic and Political Weekly*, 20 May 1971). Hitendra Desai resigned on 31 March 1971 and the governor approached Kantilal Ghia to form a Congress-R government. Ghia failed and Hitendra Desai with the support of the Swatantra and the BJS was reinstated as chief minister on 7 April 1971. In the face of continuing defection, Desai remained in power for 30 days. His advice to dissolve the assembly was initially rejected, but eventually President's Rule was declared on 13 May 1971.

After the midterm poll in 1972, Congress-R was restored with 132 seats out of 168 (Kamal 1982: 163), yet the Gujarat Congress Pradesh committee was seriously divided between rival leaders, with Ratubhai Adani, Kantilal Ghia and Chimanbhai Patel contesting for the chief ministership. In an attempt to control these and retain unity, Indira Gandhi (typically) nominated an outsider, Ghanshyam Oza. Oza was sworn in on 17 March 1972, but lacking any substantial support he proved incapable of containing any of the Congress factions within the assembly. He was ousted after 16 months when dissident MLAs within his party insisted that he face a motion of no-confidence. Oza capitulated to the Congress Parliamentary Board on 28 June 1973, as opposed to the local party committee or the assembly Congress party (Zaidi 1973: 629). The Gujarat CPP voted in a secret ballot and elected Chimanbhai Patel as chief minister on 72 votes, with his rival Kantilal Ghia securing 62. Patel was sworn in as chief minister on 19 July 1973. This assertion of the PCC's right to elect its own leader led to serious disagreements between Patel and the prime minister, and Patel's eventual dismissal.

In Bihar, the politics of intervention were equally as continuous if not more violent. Following the Congress-R landslide in the 1971 national election, the SVD coalition government in Bihar entered a protracted crisis, with the BKD unit of the coalition, led by Charan Singh, threatening to withdraw. Defection from the ministry favoured the Congress-R and on 1 July 1971 Chief Minister Karpoori Thakur resigned. He advised the governor to dissolve the assembly, but the advice was ignored and the governor called Bhola Paswan Shastri, a leader of the Progressive Democratic Front coalition that contained the Congress-R, to form the government (*Economic and Political Weekly*, 12 June 1971: 1170). This constituted the ninth government since the 1967 elections and the fourth since the states' midterm poll in 1969.

The Bhola Paswan Shastri ministry was sworn in on 2 June 1971 with a comfortable working majority. Meanwhile, Yashpal Kapoor, the prime minister's private secretary, was dispatched to negotiate a truce between two opposing Congress factions which drew on the central patronage of Jagjivan Ram and L. N. Mishra. On several occasions, infighting between these factions led to the centre dissolving the Bihar Pradesh committee, and reconvening ad hoc committees in favour of L. N. Mishra. The government was brought down when it decided to abolish a commission of enquiry (the Dutta Commission) that had been investigating corruption scandals against L. N. Mishra during his time as a state minister. In protest, the CPI withdrew with the active support of Congress MLAs loyal to Jagjivan Ram. The government collapsed after a lengthy delay in which the chief minister managed to avoid facing a no-confidence motion. State elections in 1972 returned the Congress to power, with Mrs Gandhi nominating Kedar Pande as chief minister, but infighting within the party continued and, following the advice of the prime minister, Pande eventually submitted his resignation. In his place, Mrs Gandhi nominated Abdul Ghafoor, believed to be a pro L. N. Mishra candidate, and a man with little local support (*The Statesman*, 11 December 1974; Zaidi 1972: 630).

These internal squabbles – in both Gujarat and Bihar – took place against the background of an ongoing economic crisis. In the eyes of the opposition, the government was both incompetent and profoundly corrupt. As the Congress confronted a crisis of legitimacy in the states, the peculiarities of the 'new' Congress system were crudely exposed for all to see. The appointment of Ghafoor was seen as evidence of a dysfunctional, almost pathological form of patronage. 'Ghafoor is a creature of a peculiar system by which the central leadership of the Congress select [*sic*] its nominees for the states. It is this obviously faulty system that is to blame no less than those who invented it and are protecting it whatever the costs' (*The Statesman Weekly*, 9 July 1973). Such a system, lacking any isolating institutions between state and national politics, ensured that a regional crisis of legitimacy would move swiftly to Delhi and 'the central leadership' itself, since there was no formal separation of power or responsibility to protect it.

Ghafoor proved singularly incapable of keeping his party together in Bihar. In Gujarat, Patel's Congress ministry was in open warfare with the prime minister and the centre. It was alleged, and widely believed, that Mrs Gandhi authorised the withholding of food relief to pressurise the state Congress to ditch its chief minister in Gujarat (see LSD 1974: 242). Within the state itself, Patel proved to be too close to dominant agricultural interests to run effectively the fair-price shops set up in the wake of the wholesale takeover in grains, and to even agree on any further revision on land ceilings.

In early January 1974, riots by students in Ahmedabad over a sudden increase in their mess bills quickly escalated into a ministerial crisis. Popular protests coincided with a virtual revolt within the cabinet against Patel's continuation as chief minister. Dissident ministers presented themselves to K. C. Pant, Union home minister, and demanded the centre intervene and remove Patel (*Hindustan Times*, 30 January 1974; *The Daily Telegraph*, 20 March 1974). The opposition parties, initially independent of the protests, were quick to organise anti-government feeling against Patel and the prime minister. The state-based BJS issued a communiqué to the president of India demanding the dissolution of the assembly and fresh elections: 'The economy of the country is disintegrating as a result of the collapse of the *garibi hatao* mandate.' The state government, shocked at the intensity of the rioting, took nearly two months to collapse as Patel tried in vain to assert his authority. Kuldip Nayar commented in *The Statesman* that 'people's anger in Gujarat is astounding most of the political parties there. Much of the anger seems directed at Chimanbhai Patel because he has come to epitomise the omissions and commissions of the Congress government' (*The Statesman*, 11 February 1974).

Under President's Rule, the assembly was suspended while the governor (and the prime minister's office) sought to find a new leader of the Congress, dissolving and reconvening the local Pradesh committee in the process. In retaliation, MLAs of all parties began to resign their seats in order to precipitate the need for fresh elections. Patel himself, ignoring a party directive, resigned his seat along with 44 MLAs, 23 of whom belonged to the Congress-R. In a press statement Patel said that 'Delhi had not become reconciled to my leadership but I did not expect it to carry out its politics against me on the question of food grains to such an extent as it did. How the weapon of food grains is used against me is evident from the figures of the meagre supplies to Gujarat from May last year (1973) to January this year' (*Hindustan Times*, 2 March 1974).

The speed at which the non-communist opposition was able to mobilise independent social unrest took the government by surprise. At the forefront of this was the BJS, which proved extremely effective at working with Congress traditionalists like Morarji Desai, and Gandhian Congressmen who began to converge around what would become the JP Movement. In March 1974 the prime minister specifically denounced the BJS's involvement in the agitation and accused A. B. Vajpayee, then president of the BJS, of attempting to destroy democratically elected governments. She stated that 'we know that what is happening in Gujarat is a rehearsal of what is being planned on a larger scale for the rest of India' (LSD 1974: 200). Where incidents of violence did occur, Mrs Gandhi identified the BJS and the RSS as the primary cause. In Bihar, extra-parliamentary protests began explicitly to remove the state government and to order fresh elections. Very quickly, the Bihar agitation turned into a crusade against a corrupt and 'unjust' government sitting in Delhi.

Following the success of the Gujarat agitations, the opposition parties (mainly the SSP, the Congress-O, the Socialist Party and the BJS with initial support by the CPM) agitated on 16 March 1974, attempting to prevent the joint sitting of the assembly and disrupt the Governor's Address. The agitation led to violence and rioting in and around Patna, which led J. P. to demand that he would assume the direct leadership of the movement on the express condition that participants would practice non-violent, Gandhian techniques. U. S. Dikshit and Jagjivan Ram were sent to try and ensure unity within the Ghafoor ministry and avoid the internal divisions that had ultimately brought down the Patel government. It was widely appreciated that 'if Mrs Gandhi and her colleagues fail to secure their administration in Bihar, there are fears of a widespread movement that will drag down other governments in the Union' (*The Daily Telegraph*, 20 March 1974).

On 9 April, the agitation entered its second phase wherein the non-communist opposition parties under the leadership of Narayan attempted to paralyse the working of the state government. Narayan initiated the protest in Patna with a call for 'total revolution' against a system that compelled almost everyone to be corrupt (*The Statesman Weekly*, 13 April 1974). J. P. Narayan's leadership enabled the opposition to utilise fully the economic and political failings of the Congress. As early as 1974 Narayan had charged that 'there is little or no internal democracy left in [the Congress] and its state leaders and chief ministers are mostly hand picked men, rather than leaders in their own right' (Narayan 1975: 119). In his writings on the background to the total revolution, J. P. Narayan voiced his conviction that the government no longer represented the will of the people in that 'the government has rudely disturbed the delicate balance between the executive, legislative and judiciary to concentrate more and more power in the hands of the executive, which means only the prime minister' (Narayan 1975: 121).

To the prime minister, and both her formal and informal advisors, the JP Movement was an incipient rebellion aimed at overthrowing a legitimate government that had dared to identify with the poor. The involvement of the BJS was critical to her subsequent analysis. Facing a series of no-confidence motions, Mrs Gandhi spoke out stridently against the 'loyalty' of the opposition to the idea of a democracy, and in particular, of the role the RSS was playing in organising the agitations. In a heated parliamentary exchange over the Bihar agitation, Piloo Mody (Swatantra Party) interrupted the prime minister by shouting 'What is all this sanctimonious humbug about democracy? They do not even have democracy in their own party!' (LSD 1974: 232). Combined with extra-parliamentary tactics, often Gandhian in nature, J. P. himself proved capable at bringing about a broad convergence of the opposition parties to confront Mrs Gandhi over corruption.

In August, the non-communist opposition had been unable to work together for a common candidate in the Union presidential elections (which had seen the return of Fakruddin Ali Ahmed, a close colleague of the prime minister, as Union president). Yet by November 1974, the JP Movement had brought the non-communist opposition parties into close cooperation. By November 1974, a steering committee was formed consisting of A. B. Vaipayee (BJS), Ashok Mehta and S. N. Mishra of the Congress-O, George Fernandes and Surendra Mohan (SP), Piloo Mody (Swatantra) and Rai Narain (BLD), T. Chaudary and V. Chakravarty of the Revolutionary Socialist Party, and several other representatives from regional parties such as the Akali Dal and the DMK, and ex-Gandhians such as J. B. Kripalani.

There was a paradox at the heart of the JP Movement. In the specific case of Bihar, it failed to achieve its immediate objective, and switched its attention to Delhi where it thought the chances of political success might well be greater (Wood 1975: 313–34; Dhar 2000). Yet it was within Bihar that it successfully forged a working relationship between the non-communist opposition. Furthermore, J. P.'s credentials put the Congress at a major disadvantage. Narayan's alternative, party-less democracy was either too nebulous for some Congressmen, or too closely associated with RSS doctrine to be palatable to others, but to many Congress parliamentarians, troubled by Mrs Gandhi's growing authoritarian tendencies and her failure to discipline those charged with corruption, its language struck a cord. Senior Congress politicians were concerned both by the proximity between the Congress and the communists and the particular style of Indira Gandhi's politics.

J. P.'s Gandhian socialism was, in the context of the wider ideological language used by the prime minister since 1971, a serious threat to her position. As one political commentator noted 'it is bad enough for the Congress to be on the other side of the fence as far as the JP Movement is concerned, but to align with the CPI in opposing him is unacceptable to many congressmen' (*The Statesman*, 27 December 1974). Despite her reading of the JP Movement as one 'in effect captured by the RSS and the BJS', many Congressmen called upon her to recognise it as a bastion of some of the 'best traditions of the Congress' and to open a national dialogue.

Comprehending the seriousness of the threat of the JP Movement was difficult in the charged atmosphere of 1974. The government was preoccupied with its links both to ongoing industrial action, and also its ability to undermine the legitimacy of the state governments by its continued emphasis on the complicity of the prime minister with widespread corruption. There is evidence that Mrs Gandhi was irrational in her views and attitudes to J. P. himself. Yet those that counselled a low key approach to the movement – in effect to ignore it as nothing more than a law and order issue – were right in recognising the narrow regional appeal of J. P., John Wood commented that the Movement was not broadbased, it was confined to certain areas of the state and certain sections of society and, more importantly, it was beginning to lose its way by 1975 (Shah 1977; Wood 1975: 320). Certainly the role of the RSS was exaggerated. In March of that year, Ghafoor was eased out as chief minister in Bihar and was replaced by Jaganath Mishra.

By early 1975, the JP Movement was arguably of less significance than the divisions it had brought about within the Congress over the institutional and political nature of governance, the relationship between Congress traditionalists uneasy with their association to the communists, and their disregard for an essentially Gandhian moral language that Mrs Gandhi could or would not simply understand. When delivering the Harold Laski

Lecture in Ahmedabad, Mohan Dharia reiterated his call for a 'national consensus' with the non-communist opposition over Congress policies and called for a compromise with J. P. Narayan, to reconcile government to the people. Following this address he was removed from office on the grounds that he had made inappropriate statements in public without consulting the prime minister. On 7 May 1975 in his resignation speech Dharia first read out the prime minister's letter and then went on to state that he had informed the prime minister as early as October 1974 about his concern over the presence of dubious persons within the government, and the political risks for the Congress of dealing with the CPI. He also reiterated his conviction that, while the JP Movement had been misused in particular instances by specific elements of the opposition, it nonetheless raised issues about political exclusion and inequality that should be addressed by a government committed to the abolition of poverty. He stated to the House that 'I ventured to suggest a policy of national dialogue and consensus about these burning problems. I fail to understand how this approach goes against the accepted policies of the government or the party or the basic tenets of Parliamentary Democracy' (LSD 1975: 246–47).

Widely reported in the press, the Dharia speech and the background materials it provided initiated a debate 'about the lack of a healthy system of decision-making within the council of ministers, the extent of the influence exerted by persons about the prime minister and the day to day interference in Ministerial functioning by the Prime Minster's Secretariat (*Economic and Political Weekly*, 15 March 1975: 461).

To Narayan, Desai's resignation speech revealed the utter irrelevance of democracy to Mrs Gandhi's concept of government. He called upon Jagjivan Ram and Y. B. Chavan to defend the 'true' interests of the Congress. Narayan's appeal to Ram was probably based upon his knowledge that Ram had disagreed with the prime minister's handling of the Bihar agitation and over her proximity to Railways Minister L. N. Mishra. When asked to explain why he had made such an appeal to Mrs Gandhi's senior colleagues, Narayan replied 'I am encouraging them to restore democracy within the Congress, which is only possible by challenging the absolute leadership of Mrs. Gandhi' (*Sunday Standard*, 16 April 1975).

On 27 December 1974, the political correspondent of *The Statesman* commented 'following the Gujarat and the Bihar agitations, the intangible but all important link between the Government and the Opposition has been snapped. Corruption has now emerged as the major issue' (*The Statesman*, 27 December 1975). The economic shortages that had provoked social unrest throughout India were subsiding in the early months of 1975. Although the economy was still suffering from inflationary pressures and the threat of sustained recession, prices had stabilised after reaching a peak in 1974. Stability had been brought about through short-term measures such as statutory regulations, ordinances, the use of Maintenance of Internal Security Act (MISA) and Defence of India Rules (DIR), and various anti-smuggling and hoarding legislation (*Quarterly Economic Report*, 1974: 81). Despite widespread repression, the railway strike had been brought to a close. These, coupled with hopes for a good monsoon, had increased confidence for an improved rabi crop. Significant recoveries had been registered in specific export markets such as sugar, and by April industrial production was expected to increase at a rate of 3 per cent over the year. Industrial production improved in part because of recoveries in the output of steel, coal and power (*Eastern Economist*, 11 April 1975; *Quarterly Economic Report*, 1975).

By January 1975, wholesale prices had fallen while still remaining 20 per cent higher than for the same period the previous year. In April 1975 there was a small increase and this continued into May, but it was clear that the government had managed to contain fears of galloping inflation, even if those suffering from scarcity and high prices did not immediately feel the benefit of this (*Monthly Commentary on Indian Economic Conditions*, 1975: 101; see *The Times of India*, 18 February 1975). Indeed it was the continuing perceptions of economic hardship and failure that drew attention to the political failures of the government and its slowness to act. The use of special powers to deal with anti-social activity (itself a broadening category) although effective themselves, merely drew attention to the disintegration of established forms of political activity and governance.

Charges of dictatorship – raised both by the prime minister and by J. P. Narayan – were already part of the political atmosphere of India by early 1974. As the railway strike illustrated, special legislative measures aimed at controlling the economy and preventing hoarding soon lent themselves to wider political issues such as the misuse of MISA and DIR against what many saw as legitimate political process. Congress charged the opposition with attempting to use the strike to conspire against the massive mandate of the 1971 election and a lawfully constituted government. On 9 May 1974, Indira Gandhi shouted across the Lok Sabha 'does not the opposition make a scapegoat of me for everything! Their lack of initiative, their inability to provide an alternative programme?' (LSD 1974: 533). Earlier in April, Indira Gandhi stated that the sole objective of the opposition was to remove her from the prime ministership (*Free Press Journal*, 7 April 1974).

Indeed the political situation continued to deteriorate further in early 1975 even as the economic outlook improved. On 2 January, L. N. Mishra, the central minister for railways, was assassinated at Samastipur, Bihar. Reported to be suffering from slight injuries, he was denied immediate medical treatment and allowed to continue on his journey by train. The train was subject to a series of delays and, while later undergoing an operation for metal splinters in his abdomen, Mishra died of a heart attack (The Mathew Commission of Enquiry 1975). The incident led to speculations of government – ultimately prime ministerial – involvement, while drawing attention to the deepening murk of corruption in Bihar.

The opposition charged Indira Gandhi with having got rid of an embarrassing, corrupt minister who could not be trusted to go quietly (*The Observer*, 21 January 1975). What is significant about the rhetoric surrounding the Mishra killing is that it drew both the Left and the Right into a bidding war over allegations of corruption and planned political coups. The term 'fascism' was used increasingly to identify the BJS and highlight their involvement in the JP Movement, while Mrs Gandhi was accused of being part of a communist-inspired conspiracy to declare a one-party dictatorship. The prime minister declared that the Mishra killing was nothing but a dress rehearsal for her own murder, while Jyotirmoy Basu charged that 'all the hands behind the murder are in New Delhi'. N. K. P. Salve (Congress) replied that 'an irresponsible, frustrated opposition is a menace and a real threat to the survival of Parliamentary Democracy' (LSD 1975: 339). P. G. Mavalankar concluded an urgent short-notice debate by saying that the real significance of the Mishra killing lay in that 'mystery after mystery is gathering ground. I say that this atmosphere of suspicion and doubt has been created by the mysterious functioning of the political wing of the government under the leadership of the prime minister' (LSD 1975: 439–40).

Two incidents further highlighted the atmosphere of political tension throughout India in 1975 and the level of mutual recrimination between the Congress and the non-CPI opposition. On 18 March, Indira Gandhi gave evidence at the Allahabad High Court concerning the charges of electoral malpractice filed by Raj Narain in 1971. At the hearing the police arrested one Govind Mishra in the public gallery for carrying a loaded 12-bore shotgun. On 20 March, two grenades were thrown through the window of the chief justice's car as it stopped for lights outside the Supreme Court, both of which failed to explode. The BJS charged that these events smacked of government management, aimed at creating an atmosphere conducive to maintaining the external Emergency and the use of DIR. The atmosphere throughout metropolitan India was one of anxiety and fear, with much talk of a political crisis. In an article written jointly with Arup Mallik, three months before the declaration of the Internal Emergency, Partha Chatterjee began an article with the words: 'there has been much talk in recent months about the rise of fascism in India. Some are arguing that the government, bent upon crushing the opposition by the use of "semi-fascist" methods, is turning the country into a police state' (Chatterjee and Mallik 1997: 35–58).

In March 1975, Romesh Thapar observed in the *Economic and Political Weekly* that 'there is no one in the ruling party to stop the rot. They are all frightened of the lady, and leave the job of correction to her and those around her (*Economic and Political Weekly*, 22 March 1975: 503). In reference to the apparent attempted assassination of the chief justice, Dinesh Chandra Goswami (of the BJS) warned in a parliamentary debate that 'the manner in which the attempt has been made, the earlier incident at Allahabad, the Samastipur tragedy and various other incidents involving personalities from the field of politics, lead us to the conclusion that they are not stray incidents but all part of a wider puzzle' (LSD 1975: 250). A puzzle that implied to him that the government was trying to ferment an atmosphere of fear and intimidation in preparation for increasing coercion as its legitimacy collapsed.

Yet it was to be in Gujarat, not New Delhi that the divide between the Congress and the non-CPI opposition directly manifested itself. The state had remained under President's Rule in the wake of Patel's dismissal on the grounds that drought conditions required priority to go to relief measures and not to electioneering. The non-CPI opposition argued that famine conditions had been compounded through central interference and that only a popular government could carry out relief successfully. The refusal of the Congress to specify a date for the election was 'not because of scarcity or drought, but because the Congress party in Gujarat is in a shambles. There is no leader and there is no organisation. These are the real difficulties that harness them' (LSD 1975: 377).

On 2 April 1975, Morarji Desai announced a fast-to-death to pressurise the government into calling elections in Gujarat and to curtail the continuing use of the External Emergency and MISA to parallel and supersede normal laws. As the Dharia speech had revealed the outlines of possible future rebellion within the government, Indira Gandhi was aware that the death of a man still seen as an archetype Congressman, closely related to the JP Movement and respected by the Hindu nationalist movement, would have had dangerous repercussions throughout her government. Desai's fast lasted until 13 April, and the opposition used various parliamentary procedures to keep the fast at the top of the political agenda. Ex-Gandhian J. B. Kripalani warned Indira Gandhi that if Desai died because of her refusal to hold elections violence would erupt (*The Indian Express*, 7 April 1975). On 12 April the prime minister sent a letter to Desai informing him that elections would be held on 7 June. She compromised, however, over the External Emergency, stating that, although the Emergency would not be removed, MISA would continue to be used only in the cases of 'anti-social' activities. The prime minister noted that 'I have never stood on prestige, however, I do feel that fasts of this nature are unjustified and constitute an irrational form of political pressure. Yet the prime consideration for us was the saving of Desai's life…The object of MISA has never been to curtail any legitimate activity' (LSD 1975: 320–28).

The Congress election campaign in Gujarat was organised through the prime minister's secretariat to the total exclusion of D. K. Barooah, the party president, and the state-based party organisation, which had effectively ceased to exist (interview with D. K. Barooah, New Delhi, 1985). Lists of candidates were forwarded to the prime minister after which, they were deliberated upon by the Congress Parliamentary Board. Above the immediate concerns of the Gujarat electorate, the elections were perceived as a major test of government popularity after a difficult and trying period, a tangible between regional and national legitimacy: 'The Gujarati voter has been made aware of the fact that what is at stake in the coming poll battle is not just the state's future, but the immediate future of the entire sub-continent' (*Amrita Bazar Patrika*, 8 June 1975).

The dissolution of the Gujarat assembly, despite the scale of the Congress victory in 1972 (50.6 per cent of the popular vote and 140 seats), coupled with the drama of the JP Movement in Bihar, provided a vital opportunity for the non-communist opposition at the state level. By 22 February, the former constituent parts of the Grand Alliance had elected a joint convener in the Lok Sabha (*The Statesman*, 23 February 1975). Throughout March 1975, the JP Movement continued to provide the groundwork for the coming together of the Congress-O, the BKD, the BJS and the various fragmented and often regionalised socialist parties. Only the nature of the association remained to be agreed upon, with the Congress-O and the BJS in favour of a loose federation, while the BLD favoured an out-and-out merger. On 23 April, three parties merged with the BKD to form the Bharatiya Lok Dal (BLD) (*Amrita Bazar Patrika*, 24 April 1975). On the eve of the Gujarat campaign, five non-communist parties came together to form a Janata Morcha (or popular front) and drew up a list of common candidates, prominent among them being the BJS. This degree of pre-electoral opposition unity was unusual.

The associations between the JP Movement, the RSS and the BJS were well known at the time. In 1974, Congressmen close to the CPI had issued a report in Bihar demonstrating the apparent collusion between the CIA and the RSS, and the links between this and the JP Movement noting that 'behind the façade of partyless democracy lurk the dark and violent forces of Indian fascism, well organised and well poised... the BJS, the RSS and the Anand Marg are the driving forces' (cited in Frankel 2005: 536).

In the late 1960s, J. P. had worked with the RSS in Bihar to help assist local communities suffering under drought conditions. While they disagreed on the ultimate goal of social organisation and humanitarian help, J. P. and the RSS (and through this organisation, the BJS) 'agreed on the immediate desirability of particular social welfare projects' (Jaffrelot 1996: 262). Given the degree of practical (as opposed to ideological) convergence between J. P.'s total revolution and the RSS, the organisational strength of the RSS, and the links between this and the BJS, made the JP Movement appear especially dangerous to the government, who were evidently blind to the contradictions within the movement as a whole. Writing from the perspective of the late 1990s, Jaffrelot commented that 'the JP Movement was a veritable godsend for the BJS's leaders in that it allowed them to get back into step with the [RSS] networks and integrate with the legitimate political opposition through an activist campaign outside the institutional system' (Jaffrelot 1996: 262).

Although disliking J. P.'s proclivity for using socialist references to his movement and its aims, the RSS approved of the emphasis on social uplift and the need for local organisations to redress the dominance that the state had over India under the Congress. The BJS sought to utilise the opportunities provided by this in a hope to obtain political power. In the Gujarat campaign, the BJS helped cement a popular coalition with pre-electoral seat adjustments that preserved its own identity and yet made a serious impact on the ability of Congress to retain its power in the state. On 16 May the prime minister warned that a successful win for the 'Janata' would strengthen the forces of communal chauvinism and violence, a pointed reference to the BJS, which was projecting itself as the organisational spine of opposition unity short of all out merger. On several occasions she referred to the fatal experiments carried out after the 1967 elections when unprincipled opposition alliances wrought havoc in many states. She covered personally all the 29 districts that constituted the heartlands of the opposition, specifically drawing attention to the evils of communalism. In this regard, the subsequent Congress defeat was a personal affront to her.

The Congress share of the popular vote dropped from 50.6 to 40.7 per cent. This combined with greater opposition unity to bring its total number of seats down from 140 (in an assembly of 182 seats) in 1972 to 87 seats in 1975. Congress remained the largest single party in the assembly with the Janata relying upon some independent MLAs to form a working majority. The failure of the Congress to retake Gujarat decisively implied a lack of faith in the prime minister's leadership, and the ability of the party to govern. It was, in its own right, a seismic event, and it was to be amplified by a court case verdict delivered on the same day, unseating the prime minister from her parliamentary constituency.

The Gujarat defeat became known in Delhi around noon of 12 June 1975. Later that same day, Indira Gandhi was informed by her son Rajiv that Justice Jag Mohan Lal Sinha had declared the unseating of the prime minister and her disqualification from Parliament for six years under section 8A of the Representation of the Peoples Act (Frank 2001: 371). Of the 52 charges brought against her by Raj Narain, concerning the prime minister's election to the Rae Bareilly constituency in 1971, the court had found her guilty of just two. Other more serious charges were dismissed. Yet, to many, the implications of these two events for the prime minister seemed clear. She must resign.

For the non-CPI opposition, strengthened by the Janata victory in Gujarat, the court case became the central strategy through which they sought to defeat the Congress nationally. From the very onset the non-CPI opposition brought forward moral and physical pressure to compel the prime minister to resign, aware that her resignation would bring down either an array of state governments, the central government itself, or both. On receiving the news on 12 June, the opposition parties' National Steering Committee, formed in the last week of November 1974 under the chairmanship of J. P. Narayan, called a mass non-violent protest outside the presidential residence and formulated a plan of action to force Indira Gandhi out of office. Again, in the climate of mass agitation that grew up around these events, the RSS (and the BJS) took the lead in providing volunteers for frontline action.

A. B. Vajpayee, the leader of the BJS, personally called upon the president in Srinagar to request him to call for the resignation of Indira Gandhi. On 13 June, the Steering Committee sent a telegram to the Union president stating that they would no longer recognise Indira Gandhi as leader of the Congress parliamentary party or as Prime Minister of India. The opposition organised a large meeting on the Ram Lila Grounds in Old Delhi on the evening of 22 June. At a large meeting in Calcutta on 20 June (marred by violence between the BJS and the Youth Congress), Narayan called upon the police and the army not to obey orders from a corrupt and immoral executive that no longer commanded the support of the nation.

Within the Congress parliamentary party, several prominent Congressmen, connected either with the defunct CFSA or directly with J. P. Narayan, called upon the prime minister to resign. Others, including those most intimately connected with Indira Gandhi, went directly to the prime minister's residence at Safdarjang Road as soon as the news became widely known to proclaim their loyalty. The immediate response of Indira Gandhi to the court verdict was reportedly to draft a letter of resignation to the president. The prime minister was persuaded to withhold the letter by S. S. Ray, H. R. Gokhale and Sanjay Gandhi (as well as Om Mehta and D. K. Barooah) (interviews Om Mehta, New Delhi, 1986, and D. K. Barooah, 1985) on the grounds that the charges – largely involving technicalities – carried no moral turpitude. Moreover the court had, in view of the importance of the prime minister, awarded a 20-day stay of execution to allow for appeals to be filed and alternative political arrangements to be made. Close colleagues in the cabinet and the party advised that there was no moral or constitutional reason why Indira Gandhi could not continue as prime minister until cleared by the Supreme Court. Sanjay Gandhi told reporters outside the prime minister's residence that 'there was no possibility of the prime minister resigning over such an issue' (*The Times of India*, 13 June 1975).

The prime minister telephoned the Union president in Kashmir and informed him that there was no need to return to Delhi. Once the decision to remain in office had been taken, the prime minister's secretariat reduced the implications of the court case to the simple problem of 'should the PM resign to oblige the opposition and the external enemies of the country, or should she respect the wishes of the people and remain in power?'(*Economic and Political Weekly*, 15 June 1975: 976). The decision to stay resulted in the orchestration of 'popular' support and political propaganda similar to the techniques used in 1969 at the time of the Congress split. Protests were organised by using what remained of the Delhi Pradesh committee, the influence of the Party president, the municipal resources of the city of Delhi and the patronage of the non-government offices of the lieutenant governor of New Delhi, Krishan Chad. Sanjay Gandhi and the Youth Congress were instrumental in getting the resources of Delhi behind this show of support, always acting on 'behalf' of the prime minister of India (interview P. N. Haksar, New Delhi, 1985). These protests were coordinated through the workings of the prime minister's secretariat. As in 1969, it was once more recognised in adversity that popular support was more important than constitutional propriety. The prime minister's secretariat also authorised the use of state intelligence and surveillance to monitor both the opposition's response to the court case as well as the various factions within her own party, in particular Mohan Dharia's contacts with J. P. Narayan, and Narayan's links with the BJS, especially Vajpayee and Advani (Shah Commission 1978: 20 5.27). Normal channels were also utilised to clear the prime minister through the filing of an appeal to the Supreme Court for an unconditional stay of the High Court ruling.

Indira Gandhi's appeal to the Supreme Court was drafted by her personal lawyer, edited by the minister of law and examined by S. S. Ray, the chief minister of West Bengal, himself an eminent lawyer (interview with S. S. Ray 1985). As this was prepared, formal moves were made to strengthen the prime minister's hand within the party and the government. D. K. Barooah established the loyalty of the party to the prime minister. A Congress Parliamentary Board meeting was convened in Delhi after which a formal statement was issued requesting the prime minister to stay in office. Attempts to achieve a show of legitimacy concentrated upon the Congress parliamentary party. After the statement from the CPB, D. K. Barooah, Y. B. Chavan and Jagjivan Ram made a separate statement to the press confirming their support for the prime minister (*The Hindu*, 16 June 1975). Any meaningful revolt against the prime minister's decision to stay on within the Congress would have to involve these two cabinet ministers.

On the evening of the 12 June, the prime minister summoned a meeting of the CPP executive which issued a statement that 'Indira Gandhi is a national leader and that her continuation in office is imperative'. Barooah made a statement later reiterating that the prime minister need not resign until the case had been reviewed by the Supreme Court. Yet it was already rumoured that the prime minister had decided to remain in office even in the face of specific limits being placed upon her parliamentary or constitutional activities by the Supreme Court until such time as the Court was able to pass final judgement. Such a move was technically possible because the Lok Sabha had adjourned *sine die* on 8 May 1975 at the conclusion of the Budget Session. Constitutionally there was no requirement to reconvene the Lok Sabha until mid-October, which would allow the government to escape what would otherwise be a deadlock in parliament until Indira Gandhi was cleared. Such a deadlock was inevitable given the refusal by the non-CPI opposition to recognise her as prime minister, and in doing so, cooperate with the etiquette of Parliament.

By 13 June, P. N. Haksar had prepared a document in which chief ministers, MLAs, MPs and cabinet ministers would pledge their personal loyalty to the prime minister. Many Congress chief ministers had arrived in Delhi by 13 June or sent statements to the capital confirming their support. The chief ministers of Uttar Pradesh, Andhra Pradesh, Tripura, Orissa, Karnataka, Assam, Maharashtra, Himachal Pradesh, Punjab, Madhya Pradesh and West Bengal all signed the document. In an attempt to resume the political offensive and renew the spirit of the *garibi hatao* mandate, several Congress chief ministers met in New Delhi to discuss ways in which government socio-economic policy could be implemented more efficiently. On 14 June statements were issued by most of the Congress party leaders asking the prime minister to remain in power. By 18 June, it had been possible to recall most of the Congress MPs back to Delhi to hold a formal meeting of the entire parliamentary party. Once more the potential ringleaders of any rebellion – Ram and Chavan – were used to pilot through the motion that Indira Gandhi's leadership was indispensable.

Yet fractures were already beginning to emerge within the cabinet. Swaran Singh, minister for defence, was reported to be hesitant over signing Haksar's document on the calculation that in the likelihood of Indira Gandhi's resignation, he would be called upon to be interim prime minister (*The Times of India*, 16 May 1975). Earlier, Jagjivan Ram had been publicly forced to deny that he had sent the prime minister a letter suggesting that she should resign and staking his claim to the office (*The Times of India*, 17 June 1975). This followed a speech in which he had said that the legitimacy of the courts must be upheld (*The Statesman*, 21 June 1975). It was widely known that Jagjivan Ram had a following of between 60 and 80 MPs in the parliamentary party and that these could be used to prevent Indira Gandhi nominating an interim leader of the Congress parliamentary party. Mohan Dharia had been warned beforehand not to raise any objections, and he and Chandrasekhar, Ram Dhan, Krishna Kant, Sambhu Nath Mishra and Lakshmikanthamma, all former members of the CFSA, stayed away. Chandrasekhar stayed away from subsequent meetings of the CWC. Dharia alone argued publicly that the prime minister should resign (interview Mohan Dharia, Pune, 1986).

On 20 June the prime minister moved the Supreme Court to fix a date – 23 June – to record a petition of appeal for an absolute stay on the Allahabad verdict. The prime minister's lawyer argued that the granting of a conditional stay would damage the moral position of the prime minister even if it did not damage her constitutional position. On 24 June the Supreme Court ruled that the prime minister could have a conditional stay from the Allahabad Ruling until the case could be either confirmed or quashed. The conditions were that 'She will neither take part in the proceedings of the Lok Sabha, nor vote, nor draw any payments in her capacity as a member of the Lok Sabha' (*The Times of India*, 25 June 1975).

Justice Iyer commented that the conditions substantially preserved the position of the prime minister. Given that Parliament was not in session the disqualifications on voting and drawing pay were deemed to be largely academic and irrelevant. Yet this implied to the opposition that the Lok Sabha would be kept adjourned until the Supreme Court had deliberated, a move calculated to alarm them. The Indian Constitution required parliament to meet twice a year (Thakur 1995: 143) and given the probably lengthy appeals process, continual suspensions of the parliament would not prove politically expedient.

The Congress Parliamentary Board used the judge's statement to argue that their earlier position had been vindicated, a statement that was once more individually supported by Jagjivan Ram and Y. B. Chavan. The terms of the stay seemed, however, to reinforce the sense that key elements within the national party were losing confidence in Indira Gandhi's moral position. Following the announcement, Congress rebels associated with Chandrasekhar met openly for the first time to discuss the damage being caused to the government and to the office of the prime minister by her refusal to resign (*Hindustan Times*, 25 June 1975). Following the awarding of the conditional stay the opposition went ahead with a huge rally on the evening of 25 June at which Narayan once more called upon the police and the army to disobey 'immoral' orders. He also called upon Justice Ray not to sit on the Supreme Court bench that would consider the prime minister's case on the grounds that he was politically biased following his controversial appointment by Indira Gandhi herself.

Narayan's call to the army to 'rebel' was the main excuse given by the government to justify subsequent events in India. Yet the sequence of events throughout India that would dovetail into the Emergency substantially predate the rally of 25 June and coincided rather with the fears associated with the anticipated implications of a conditional stay order from the Supreme Court. As a turning point, this conditional stay is almost as significant as the original Allahabad Verdict. (Shah Commission 1978: 21 5.29). The second ruling came at a time when the national fortunes of the non-communist opposition were still undecided. The CPI took the line that the JP Movement was reactionary, underpinned by both a reactionary bourgeois but also communal ideology. As such Congress must be supported even if it had failed generally to implement its own economic programmes and was itself internally fragmented between progressive and reactionary elements. Yet the enthusiasm of the opposition to seize the political initiative in the run-up to the appeal and the resulting judgement alarmed the prime minister, her advisors and personally nominated chief ministers and party presidents. Justice Iyer's ruling raised the question: could the government cope with the situation that was now likely to develop if Indira Gandhi was to remain in office? The level of opposition activity, and her own concerns at events within her own party, implied to her that the issue of the court case could not be trivialised. It implied that the longer the political uncertainty continued, the harder it would be to retain control over the Congress. It became clear that Indira Gandhi could retain the prime ministership only if she could eliminate the opposition's challenge, restrict the channels of permissible protest, undo the court case without having to undermine the legitimacy of the Courts and secure beyond question the legal position of the prime ministership (Shah Commission 1978: 21 5.31).

On 24 June, informed about the opposition's rally planned for the following evening, the prime minister contacted various chief ministers and requested them to make themselves available in Delhi for the evening of 25 June and to prepare their respective states for the detention of opposition leaders. By the afternoon of 24 June, the chief ministers of Andhra Pradesh, Karnataka, Madhya Pradesh, Rajasthan, Punjab and Bihar had been taken into confidence concerning what was planned and asked to make preparations (Shah Commission 1978: 23 5.46).

Earlier on 25 June a meeting had been held in the prime minister's office, South Block, New Delhi, with R. K. Dhawan (private secretary to the prime minister), Om Mehta, Bansi Lal and the heads of the CID, the Intelligence Bureau and the Police Commissioner (Shah Commission 1978: 22 5.42). Just before lunch, S. S. Ray was brought to the prime minister's house in his capacity as confidential friend and lawyer. Indira Gandhi allegedly asked him if there was any way that she could constitutionally detain the opposition leaders and 'save India'. Ray stated that, while they were discussing the court case, a bearer came in with a slip of paper confirming news about an opposition call to the army to rebel. Ray also noted to several sources that the prime minister showed him intelligence reports citing external support for the JP Movement, explicit links to a RSS–BJP strategy to undermine her, including possible funding and involvement from the CIA. The Intelligence Bureau had reported that the main organiser of the forthcoming rally on the Ramlila Grounds was a member of the RSS, one Nanaji Deshmukh, who was working closely with Desai to bring her down (Austin 1999: 304). It was reported that J. P. would call on the army and the police to rebel against the government, as well as calling on people to withhold their tax from the state. Summarising his evidence before the commission, the Shah report stated:

> She [Indira Gandhi] had told him on two or three occasions prior to this that India required a shock treatment and some sort of emergent power or drastic power was necessary. He mentioned that one of the occasions when she had mentioned shock treatment had been before the Allahabad Court Case.
>
> (Shah Commission 1978: 21 5.41)

S. S. Ray left Safdarjang Road to examine the relevant parts of the Constitution. He returned at about 4.30 p.m. briefed with the relevant clauses, and a document summarising previous cases in the United States where the US Supreme Court had sanctioned the use of 'extraordinary legislation' in defence of a legitimately constituted government. He suggested that the Prime Minister should declare a State of Internal Emergency under Article 352(ii).

5 The state as political impasse 1975–1980

The sequence of events reveals that arrangements for the detention of the opposition were made before it was decided under what legislation the detentions were to be carried out. Since S. S. Ray was consulted two days after the meeting at the prime minister's house, it can be presumed that the arrangements that were discussed on 23 June were to be carried out under the then existing laws of MISA and DIR. At one point in the deliberations between 23 June and S. S. Ray's involvement on 25 June, it was decided that the existing powers of the government, although adequate under conditions of an 'external' Emergency, would not sufficiently justify executive action within India against the opposition parties. It was paramount to the prime minister that the Emergency appeared to be a legitimate act.

After S. S. Ray's suggestion concerning an internal emergency, Indira Gandhi asked him to accompany her to see the president. At the first meeting with Fakhruddin Ali Ahmed, Indira Gandhi mentioned her decision to call for an internal emergency. Three questions were raised with the president: Could a decision be made without consulting the cabinet? How should the proclamation be worded? In what language should the proclamation be written? On returning from the president's house, S. S. Ray advised the prime minister to consult on these matters with Barooah and 'involve the other leaders' (Shah Commission 1978: 23 5.47). He also stated that under a procedural process outlined in the Government of India (Transactions of Business) Rules, the prime minister could act to seek cabinet approval retrospectively.

On her return to Safdarjang Road, Indira Gandhi, D. K. Barooah and S. S. Ray proceeded to draft the letter to the president. When later testifying to the Shah Commission S. S. Ray recalled how this meeting had been interrupted continually by Sanjay Gandhi, who would insist on talking to his mother in private. After the letter had been drafted, S. S. Ray learned from Om Mehta that plans were being formulated to 'close down' the high courts and disrupt the electricity supplies to the press: 'Shri Ray was surprised because he had told her [the prime minister] that under the Emergency one could not take any action unless rules were framed. Shri Ray said that locking up the high courts and cutting off the electricity connections could not just happen' (Shah Commission 1978: 24 5.47). He was allegedly approached by Sanjay Gandhi, who accused him of 'not knowing how to run the country'. It was only at this stage of the proceedings that the home minister – Brahmananda Reddy – was informed about the plans for the Internal Emergency. Detentions had been arranged through Om Mehta.

Reddy was reportedly shocked at the possibility of two states of emergency existing at the same time and suggested that the moves being considered should be carried out under the External Emergency (Shah Commission 1978: 24 5.49). Reddy was then shown a draft of the letter that was to be delivered to the president. Meanwhile S. S. Ray had scrutinised the Business Rules and Procedures of Government and discovered that, under the 3rd Schedule, the prime minister could take a decision without consulting the cabinet so long as such a decision was ratified within six hours. The letter was delivered to the president at about 11.30 p.m. by Shri Balachandran, secretary to the president. It was rejected on the technical grounds that it placed undue emphasis on the term 'personal satisfaction'. As stated in the Shah Commission, 'the letter from the prime minister indicated that the cabinet had not considered the matter. It was worded in such a manner as would make it appear that the decision to declare the Emergency was of the president' (Shah Commission 1978: 25 5.51).

Ali Ahmed pointed out that, since the prime minister was constitutionally bound to consult the cabinet, the 'personal satisfaction' of the president did not arise. The letter was subsequently redrafted and submitted to the president just after midnight (interview with S. S. Ray, 1986).

Substantial evidence exists to show that opposition leaders were being taken into detention throughout northern India before the legal niceties with the president were concluded. Various members of the government who had until now been totally unaware of the declaration were gradually informed. P. N. Dhar, the prime minister's secretary, came to know what was happening around 11.30 p.m., when he was handed a text of the speech Indira Gandhi was to make on All India Radio (AIR) the following morning. B. D. Pande, the cabinet secretary, was enlightened only when he was directed to organise an important cabinet meeting at 6.00 a.m. that morning. (Dhar 2000: 123). H. R. Gokhale, as well as Y. B. Chavan and Jagjivan Ram, were actually informed at the cabinet meeting itself (interview with Bahuguna, New Delhi, 1986). As the Shah Commission was to summarise two years later, 'on the basis of the evidence it is clear that some of the important functionaries in the Home Ministry, the Cabinet Secretariat and the prime minister's Secretariat, who should have been consulted before such an important decision was taken, did not know about the Emergency until very late, some as late as the morning of 26th June' (Shah Commission 1978: 25 5.53).

In Delhi arrests were carried on through the office of the lieutenant governor following direct orders from the prime minister. It has been alleged but never proven that Sanjay Gandhi and R. K. Dhawan also assisted in preparing detention lists. Throughout northern India, 1,012 people were detained under MISA in one night. (Home Ministry Report 1974–1975: 39–134). While the MISA regulations empowered only the lieutenant governor and men of corresponding rank to enact preventive detention, the size of the operation required this power to be delegated to the various functionaries and civil servants directing the arrests. The Defence of India Rules (1971) were amended under a presidential ordinance on 30 June 1975 and were renamed the Defence and Internal Security of India Act. This amendment enabled district magistrates and commissioners of police to exercise powers that had hitherto remained at the disposal of the state governments alone. Such devolution of power also involved junior police officers that were issued blank MISA permits through the lieutenant governor. Out of Delhi, this delegating of authority devolved down to the district magistrates and the district collectors (Shah Commission 1978: 39–134).

The prime minister was aware that, since the press would protest over the detention of opposition MPs and MLAs and alarm urban middle class opinion, it was important that press censorship coincide with the carrying out of mass arrests. The government prevented the publication of the Delhi newspapers on 26 June by putting pressure on the lieutenant governor to disconnect the electricity supply to the main press street in New Delhi (Shah Commission 1978: 22–23). Instructions for this disconnection had been conveyed orally to the lieutenant governor on the evening of 25 June who had then contacted the Delhi Electrical Supply Undertaking. In the three days it took the government to set up the required apparatus for censoring the press, electricity supplies to the press houses were continually interrupted (Janata Government White Paper 1977).

V. C. Shukla – by then minister of information and broadcasting – had met the prime minister on 27 June to discuss the need to establish personal control of the media and had immediately set up a Coordinating Committee consisting of editors, media representatives and senior civil servants that met throughout the Emergency (Janata Government White Paper 1977: Chapter 2). By September 1975 a whole series of constitutional and statutory enactments had further undermined the independence of the press without any visible outrage. Concerning the need for censorship in general, the home ministry had innocuously commented that 'the basic object of censorship is to prevent the publication of a distorted version of any incident which might have the effect of causing alarm among the general public or create disaffection towards a lawfully constituted government' (Home Ministry Report 1975–1976).

The Prevention of the Publication of Objectionable Matter Ordinance, passed in August, further eroded the ability of the press to report proceedings in parliament as well as going a considerable way to setting up self-censorship. The powers conferred on the government were draconian in that objectionable matter was defined as any written matter or article 'likely to incite hatred against the government established by law in India or in any other state'. The Press Council Amendment Ordinance abolished the Press Council and set up a news agency under the auspices of the government. The news agency Samachar was created by the forced merger between The Press Trust of India (PTI) and the United News of India (UNI). H. N. Mukherjee (CPI) argued in the Lok Sabha that the introduction of these three ordinances alone scuttled the freedom of information as established throughout India (LSD 1976: 121).

There were no notable protests over the measures taken by the government, and no widespread demonstrations. The leadership of the opposition had been effectively isolated. No political parties were banned, with the exception of the RSS and the Jamaat-I-Islami party. The RSS claimed that over 25,000 social activists were arrested at the beginning of the Emergency, rising to over 80,000 by the end of the first year. The leadership would later claim that the government had especially targeted them. Commenting on the consequences of the event, Jaffrelot notes that the proclamation of the state of emergency would allow the Hindu nationalists to strengthen their organisation within the opposition, precisely because of the strength of their network of activists, who were more capable than many others of standing the test of clandestine action and organising the satyagrahas (Jaffrelot 1996: 277–79).

By 30 June, all the policies and statutes relating to the External Emergency had been quickly recodified to be relevant within the new context of the Internal Emergency. Further amendments to MISA after 25 June involved changes that allowed for the detention of foreigners and procedural changes that allowed the grounds on which detention had been awarded to be treated as confidential matters of state. As such they could not be disclosed to the detainee or even the advisory board but only the arresting agent. Later amendments meant that clearance on one charge of MISA did not prevent immediate re-arrest on another charge.

Following these immediate measures, the Congress began a process of consolidation that involved bringing into action the normal institutions of parliamentary democracy, such as the cabinet, the Parliament and various non-governmental organisations, through which to normalise the Emergency ordinances and to reassure both domestic and foreign commentators that the political system would remain essentially the same. On the morning of 26 June, at 6.00 a.m., the Indian cabinet met and discussed the Internal Emergency for the first time. It discussed the various measures that had been carried out during the night and decided that the emergency declaration should be followed by announcements of radical socio-economic reforms. The cabinet met again that evening and discussed matters relating to arrests and detentions. The cabinet was informed by the prime minister that only 15 per cent of all the arrests were political (i.e., carried out under MISA), the rest being equally divided between offences under COFEPSA (Conservation of Foreign Exchange and Prevention of Smuggling Act) and the IPC code (*Hindustan Times*, 29 June 1975). The official figure for non-CPI opposition leaders taken into detention was 876 (Ministry of Home Report 1974–1975). The prime minister announced the setting up of an Emergency Committee consisting of Y. B. Chavan, Jagjivan Ram, B. K. Reddy, H. R. Gokhale and Om Mehta. The Emergency Committee duplicated the earlier Political Affairs Committee, with the significant add-ition of Om Mehta. Throughout the first two months of the Emergency, the cabinet met 28 times, while the Emergency Committee met on 41 occasions (*Hindustan Times*, 4 October 1975).

The process of normalising Emergency statutes began with the Monsoon session of the Lok Sabha, which had been called to session on 21 July 1975. Although closing down Parliament was probably not practicable, and would have shocked many supporters of the Emergency as being counter-productive, the government altered the procedures under which the Lok Sabha operated. The government was well aware that previous difficulties had been brought about, not through the number of opposition MPs but in their ability to manipulate parliamentary procedures. On 21 July, Minister for Works, Housing and Parliamentary Affairs Raghu Ramaiah moved that during the current session, due to extraordinary circumstances 'all other business, including Calling Attention Questions and Private Members' Bills, stand suspended until further notice' (LSD 1975: 226).

Mohan Dharia criticised the move as nothing more than 'the virtual sur-render of the Sovereign Parliament to the Executive'. Such a move assisted the government in voting through the Emergency proclamation into law and also passing through a record number of bills. By June 1976, three ses-sions of the Lok Sabha had passed more than 100 bills while the president had promulgated 34 ordinances (*The Times of India*, 10 July 1976).

Immediate amendments concerned the consolidation of the political goals of the Emergency. The 38th Amendment Bill related to the president's 'satis-faction' on matters of ordinances, and made this satisfaction non-judiciable. This in effect placed the ordinance powers of the president, 'Administrators and Governors' beyond legal scrutiny. The 39th Amendment related gen-erally to the privileges of the president, vice-president, prime minister and speaker of the house, and placed them all above judicial scrutiny. It also sought to 'render pending proceedings in respect of such elections under existing laws, null and void'. The newly inserted Article 329(A) established a special parliamentary forum to investigate charges of electoral malpractice and corruption against the president, the vice-president, the prime minister and the speaker. When introducing the 39th Amendment, H.R. Gokhale stated – evidently without irony – in the Lok Sabha that 'it is ridiculous that the prime minister, recognised throughout the length and breadth of the country as the undisputed leader, should be subjected to a process in which judicial scrutiny takes place' (LSD 1975: 52).

The 38th Amendment to the constitution was introduced on 23 July, and by 28 July had been ratified by 15 states. On 7 August the 39th

Amendment was formally passed by a vote in the Lok Sabha. In the meantime, the Supreme Court had continued with the hearing of the Allahabad appeal despite the fact that the laws under which it had been constituted had been amended, and that the grounds of the original charge were no longer deemed criminal. When the Supreme Court quashed the Allahabad verdict in November 1975 it did so by upholding the validity of the 39th Amendment, not by clearing Mrs Gandhi of the charges themselves. Further ordinances issued throughout the regime sought to pressurise the judiciary, and to remove it as an obstacle not just to 'social reform' but also to the style of government that Mrs Gandhi appeared to favour. The presidential ordinances relating to the actual declaration of the Internal Emergency (352 and 359) had suspended the ability of the courts to move in defence of the fundamental rights conferred by Articles 14, 19, 21 and 22 of the constitution. A separate ordinance was issued on 8 January 1976 suspending any citizen from moving the courts in defence of the fundamental rights listed in Article 19 of the constitution (Ministry of Home Report 1975–1976). Despite this, and regardless of subsequent ordinances, the high courts continued to submit writs in defence of human rights against MISA. By August 1975, nine high courts had allowed writs of habeas corpus to be admitted in spite of Article 359 suspending their right to do so, on the argument that common law could not be suspended through a constitutional device.

Having clarified the statutory basis of the Emergency, Mrs Gandhi quickly sought to consolidate her support within the cabinet, in effect by rewarding those who had stood by her and punishing those who had shown any signs of diffidence. On 28 June, I. K. Gujral was removed from the now critical information and broadcasting ministry because he had shown obvious unhappiness at the way in which the press was being censored (interview with I. K. Gujral, New Delhi, 1986) especially with regard to jamming foreign broadcasts. He was replaced by V. C. Shukla (interviews with P. N. Haksar and V. C. Shukla, 1985). At the end of November, G. S. Dhillon, the former speaker of the Lok Sabha, became minister of transport and shipping, Bansi Lai initially became minister without portfolio, and finally defence minister. Swaran Singh and U. S. Dikshit were both dropped (*The National Herald*, 1 December 1975).

The emergency provided the context for the prime minister to press on with the various institutional changes to central government that had featured since the early 1970s. Not surprisingly, it provided a further rationality for centralising decision-making within the senior ministries of home and finance, and the prime minister's secretariat. The finance ministry under K. C. Subramaniam was formally divided, with income tax and revenues being placed under the concerns of Pranab Mukerjee. A close associate of Sanjay Gandhi, Mukerjee was given direct access to the prime minister (see Nayar 1977). Through the earlier extension of detention ordinances relating to anti-smuggling activities (under COFEPSA) and the increased powers of tax collectors and customs officers generally, Mukerjee controlled a very important department that was used widely to intimidate people on charges of tax evasion and fraud. Such legal actions were pursued without any consultation with the law minister.

Significant areas of policy moved from ministry to ministry, with the continued denuding of the powers of the home ministry. For example, the prime minister initially took personal control of press censorship before it being handed back to the home ministry, but with whole areas transferred to the prime ministers secretariat. The ministry of home affairs was continuously pointing out to the prime minister (and the law ministry) that it had no precise idea who was being detained under what legislation. In particular, permission to use MISA against journalists came from the prime minister, ignoring the procedures for monitoring detention cases that lay with the home ministry, and often (as it turned out) without any written authorisation. Within the home ministry in particular, Om Mehta, enjoying a personal friendship with the prime minister, the prime minister's son and in command of the powerful department of personnel affairs, was working independently of his superior in the cabinet. Many of the policy responsibilities of the home ministry throughout northern India and the Delhi Union Territory were implemented through the DMC (Delhi Municipal Corporation) under the control of Sanjay Gandhi and Krishan Chad. Only in March 1977 did the home minister 'resume charge of his department' (Sahgal 1982: 117).

Again, although open to an ideological justification, such centralisation and personalisation of power was largely dysfunctional. By the beginning of 1976, the prime minister's secretariat had extended throughout the entire government, using 'special assistants' who reported to R. K. Dhawan and the prime minister directly (Shah Commission 1978: 230 24.14). Virtually all these special assistants were senior civil servants of the IAS S. K. Misra (defence), N. K. Singh (commerce) and V. S. Tripathi (information and broadcasting) all worked with their relevant ministers, ministers of state and directly with the prime minister but in an entirely unstructured and reactive manner (Ministry of Home Report 1976–1977). Ministers, aware that Indira Gandhi had alternative ways of monitoring their departments and their attitudes towards the Emergency, consequently distanced themselves from suspect bureaucrats who protested of external and political interference. This seriously affected the responsibilities of the bureaucracy, which found itself operating under conflicting loyalties. Again, although it is difficult in 2006 to clarify the logic as intentional, there is every indication that such a process, such an idea of government, was almost deliberate.

As the dust settled from the immediate priorities of the Emergency, such as detentions, censorship and the undoing of the court case, there appeared to be genuine confusion as to what the aims of the 'new era' should now be. As an overview of the events between the 12–24 June 1975 reveals, Mrs Gandhi went to considerable lengths to ensure that the declaration of the Emergency was both legal and convincing at all levels: nationally, internationally and, perhaps most important of all, to herself.

From 26 June to 1 July 1975, the government sought to defend the Emergency by reference to the opposition's attempts to incite rebellion and to work against the unity of India (Rao and Rao 1975). After July, the justification became the need to restart the radical socio-economic reform initiated by the 1971 *garibi hatao* mandate and renew the government's socialist credentials. Such a justification was implied in the very first prime ministerial broadcast on 26 June. On 1 July, the prime minister announced a 20-point programme and called for clear vision and determination to ensure successful implementation of the new economic program. In effect, the state sought to open up channels to key sections of society – notably the middle classes and middle peasants – to enhance cooperation and collaboration with the state, as well as to ensure a more efficient form of resource extraction. It tried to do this without visibly abandoning the scheduled and backwards castes that had been directly mobilised behind the 1971 elections. The new programs touched upon issues such as price controls, foreign reserves, taxation and land reforms, 'some of them are new. Others were set forth earlier but require to be pursued with greater vigour and determination than has earlier been possible' (Ministry of Information and Broadcasting 1975: 5).

On the industrial front, the prime minister had commented on 27 June that 'wild conjectures are circulating about the impending nationalisation of industries and new controls. We have no such plans. Our purpose is to increase production and output' (Ministry of Information and Broadcasting 1975: 6). On 28 July 1975, the Lok Sabha passed the Taxation Laws (Amendment), which prescribed stiffer penalties for black money operators and those found guilty of fraud. The bill increased the mandatory powers of income tax collectors who were now under the supervision of the prime minister's secretariat. Other bills introduced included the Payment of Bonuses, Regional Rural Banks, Import and Export, and the Voluntary Disclosure of Income and Wealth.

Emphasis on bureaucratic efficiency allowed the public sector to utilise excess capacity and achieve sudden increases in output, even in cases where demand was insufficient to clear excess stocks (Toye 1976). Ordinances issued before the Emergency relating to tax evasion and controls on foreign exchanges were merely added to and implemented more rigorously. Economic management in the public sector continued to be marked by pragmatic attempts at increasing output through a streamlining of industrial licensing. The prime minister had stated in her broadcast that 'licensing procedures had come in the way of new investment and would be amended accordingly'. Anti-monopoly policies would be allowed to fall into misuse where they frustrated specific companies from extending their capacity, especially in relationship to export.

As early as August 1975, inflation had been virtually eliminated within the domestic economy, and news about a good harvest was beginning to affect the psychology of the grain markets. This led to a decrease in wholesale prices and a small increase in per capita incomes. By the end of the 1975 rabi season, agricultural output had yielded an estimated 118 million tonnes of grain and had led to rumours that India would be exporting grain to the world market. The wholesale price index for agricultural goods in July 1976 had moved down to 336.6 per cent, with cereals at 305.6 per cent and pulses down to 303 per cent (*Eastern Economist*, 31 October 1975: 829). Again these improvements occurred in the absence of any structural change. Likewise in industry, initial estimates in July 1976 indicated that output would be about 10 per cent up over the total of 1975. A significant contribution to this was improved power supplies, which once more reflected the success of the monsoons. More indirect factors behind the economic recovery involved the increase in remittances from Indians living abroad, especially in the Gulf. Significantly there had also been increases in foreign aid through the workings of the AID India consortium and the International Monetary Fund (*Eastern Economist*, 9 August 1975; *Quarterly Economic Report* 1976: 22).

In a situation of general economic recovery, the 1976 budget continued with the moderately reflationary package outlined the previous year. It stepped up the investment outlays for the public sector, although further cuts in direct taxation led to a necessary increase in the level of indirect taxes. Increased demand and availability of non-essential consumer goods led to further modest growth. To a large extent this flexibility had been achieved through the adoption of an annual plan within the overall framework of the fourth five year plan. This increase in growth was significant. In a recent review of the Indian economy from 1950 to 1980, one author has pointed out that the pickup in growth of Indian manufacturing between 1975 and 1980 'is to be attributed to the performance of industry during 1976' (Balasubramanyam 1984: 136).

The Congress was understandably eager to take credit for the economic gains of the Emergency. In the first few months they acted to cement a level of goodwill and cooperation between the regime and metropolitan middle classes as well as key constituencies throughout rural India that had been marginalised through scarcity, the chaos of the wholesale takeover of wheat and the collapse of the fair-price distribution shops. By December 1975, much was made of the government's success in having curtailed inflation and in having limited the growth of the money supply to 6.3 per cent in 1975–1976, compared to the 16 per cent expansion for 1974–1975. At first glance these gains were impressive, yet even at the time there was a sense that the convergence was in part coincidental. The initial return to stability within the Indian economy had been revealing itself since early 1975 after two years of serious recession and inflation (*Quarterly Economic Report*, 1974). Improvements within the domestic economy did, in a majority of cases, significantly predate the Emergency declaration. This was partly in response to a good monsoon as well as strict fiscal and monetary controls by the government. Economically the Internal Emergency merely furthered a policy of cautionary reflation after three years of deflationary policies.

Although it is difficult to assess the specific effects of coercive measures such as MISA and anti-smuggling legislation, they had been used increasingly after 1971 to prevent upward pressure on wages through strikes and disruptions. During the Emergency, fear ensured a decrease in the loss of man-days due to strikes and labour disruptions. In August 1975 the number of man-days lost through industrial action had dropped to 16,127, compared with 136,000 for the same period in 1974 (*Hindustan Times*, 15 October 1975). By January 1976, the powers granted to the tax collectors and the police in combination with a Voluntary Income Disclosure Scheme had allowed the government to uncover INR750 crores of unpaid tax. Giving a statement in the Lok Sabha, the finance minister stated that the income tax authorities had seized between 1 June 1975 and 30 November 1975 INR1280,64 lakhs in 1,345 separate raids in metropolitan India (LSD 1976: 101). Yet by 1976 the Indian government was actually concerned over the excessive shedding of labour by large corporations and multinational subsidiaries. By January 1976, 35,000 workers had been dropped from private sector undertakings, 18,000 of these had been shed through lockouts, with the remaining 13,000 being laid off. In February 1976 the minister of labour introduced into the Lok Sabha the Industrial Disputes (Amendment) Act in an attempt to curb the increases in 'unjustified' layoffs (LSD 1976: 71–83). The bill sought to make government approval necessary for layoffs by firms employing more than 300 people, a statutory requirement that was ignored as much as evaded.

In November, the labour minister confessed in a written answer in the Lok Sabha that layoffs and retrenchments were continuing unabated and that the government could not prevent private companies taking advantage of the Emergency to restructure themselves. Such confessions fuelled concerns that an unstructured liberalisation of the economy would quickly undermine India's legacy of planned economic growth and import substitution. Such blatant free market moves were increasingly criticised by the CPI, and began to open up the ideological justifications for the Emergency to closer and greater critical scrutiny by those who had possibly supported it the most. From late 1975, the CPI increasingly urged the Congress to use the Emergency to radicalise the economy in line with government economic policies. Instead the Emergency was used to encourage foreign industrial corporations to enter the Indian economy and dismantle government regulations within the state sector that appeared to be holding back production. Again, this was not so much a decisive shift in strategy underlying the class interest of state action, but the easiest option by the state to take when confronted with powerful societal interests it did not seem able to either influence or intimidate.

This pattern was also prevalent throughout the countryside. Disguised by a reassertion of socialist and reformist zeal, the state nonetheless increasingly abdicated reform to the very social forces disinterested or actively opposed to carrying it out. The problem lay with the failure of the Congress, in the wake of the split, to mobilise the poor into active party members, and thereby reproducing a cadre-based structure that could coordinate the implementation of policy. As such, the Emergency did not enable the Congress to make significant strides on matters of land reform or agrarian policy in general. Despite conditions of excess agricultural production and coercive power, the government decided against stepping up the procurement of grains for its public distribution system and did not press ahead with fresh initiatives against the wholesale grain sector, despite the call from the Left (and the CPI) in particular to do exactly that. Despite the size of the 1974–1975 harvest, wheat and rice procurement levels were lower than those in 1973–1974. In September 1975, the *Economic and Political Weekly* argued that 'the opportunity provided by the Emergency must be utilised to recoup more of the Kharif [summer] crop than has normally been possible. The centre could, under actual conditions, legislate against hoarders and make procurement prices compulsory' (*Economic and Political* Weekly, 2 August 1975: 1140).

Such arguments exaggerated the ability of the government to act decisively in specific policy areas as opposed to merely legislate. As it was, the support prices of grains did not fall. J. F. J. Toye notes that procurement prices of wheat remained at INR105 a quintal despite a fall in the market prices of 25 per cent a quintal, a strategy that gave resources back to dominant agrarian interests (Toye 1976). Any distribution of land that occurred during the Emergency was in many cases related to earlier legislation carried out before the declaration (*Economic and Political Weekly*, 13 March 1976: 413; Omvedt 1985). Following national guidelines laid down in 1972, 2.16 lakhs had been declared as surplus (under various ceiling laws) and 19,000 acres had actually been redistributed by early 1975. Following the political events of 1975, only an additional 6.5 lakhs of an estimated 40 lakhs surplus was declared. By May 1976 only 1.3 lakhs had passed into the hands of the landless. Although Emergency legislation prevented undue litigation holding up the distribution, the government made no attempt to alter the exemptions to the ceiling laws that allowed landlords to evict cultivators on the grounds that they were bringing the land into 'personal' cultivation.

What is significant in drawing attention to declining state capacity is that various attempts were made to reconstitute the necessary institutional structures and the wider societal collaboration needed to push on with the reform process. It was as if the Emergency created a peculiar space in which debate and policies replaced social movements and implementation. The chief ministers had met in New Delhi in March 1976 to discuss the implementation of land-reform policies. They had decided on 30 June as the deadline for the carrying out of land acquisitions and redistribution. Several reports concerning the 'success' of rural reform appeared in the press, especially in Bihar, where under the Land Ceiling Act 31,000 acres had been notified as surplus. Yet the agricultural ministry's review in May criticised the lack of political will and argued that programmes were being stalled through the lack of popular participation. By July 1976 the commitment to carrying out land reform had been replaced by a vague concern about productivity levels.

Following the first cabinet meeting on 26 June, D. K. Barooah issued a party circular at the end of 1975 to all chief ministers and party presidents concerning moves to set up the machinery through which the 20-point programme would be implemented (Zaidi 1976: 159). This followed an initial circular dated 28 August 1975 that concerned party policy and guidelines for mobilising support. Each state was instructed to establish a three-tier system, with each level having responsibilities towards specific aspects of uplift and reform. The circular had emphasised the need for the Pradesh, district and taluk committees to cooperate with the panchayats over involving the rural population in economic development. Barooah had set aside 26 June 1976 as Discipline Day, and the following seven days as Dawn of the New Era week, the irony of such fascist-sounding titles apparently lost on him, as well as the Congress-R as a whole, if appreciated by some elements of the press.

By 26 July 1976, a seventh circular had been issued that drew attention to the reported inactivity of the district Congress committees which stated that 'the District and subordinate Congress Committees have to play their full role in creating a climate suitable to the implementation of the 20-Point Programme. It is therefore imperative that the district Congress Committees should start functioning effectively' (Zaidi 1983: vol. 24 159).

D. K. Barooah set up a seven-man committee to discuss the ways in which the Congress could convert itself into a cadre party and improve on the implementation of government policy. Again the ghost-in-the-machine here was the RSS and the successful ideological base that the RSS gave to the specifically political organisations like the BJS. Although one was banned and the other locked up, the Congress-R sought to emulate them, and in particular encouraged the Youth Congress to become a sort of parody of a vigorous, dynamic movement that would rejuvenate the parent body. Yet serious concerns were expressed about the conditions of the Pradesh committees, and the continuing proliferation of ad hoc committees set up under the auspices of the centre, which had no social support and where they did, did not represent the social constituencies necessary to carry out a radical redistribution of resources.

An opportunity to re-engage with the earlier social radicalism of the pre-Emergency period suggested itself with reference to property rights, and what would become the 42nd Amendment. The decision to undertake constitutional amendment arose in part through the government's obsession with the courts and the fear that the legal system had become the bulwark of 'reactionary, propertied interests against a radically elected government'. Did not the Emergency provide the opportunity to solve outstanding controversies concerning Parliament's right to amend the constitution in line with government economic policy, issues that had bedevilled the government since 1969, and even further back, to the time of land-ceiling legislation in the 1950s? In seeking to solve the dispute between the directive principles and the fundamental rights, it was argued that the Emergency could lay the legal foundations for radical economic and social reform. Such an attitude was consistent with earlier sentiments that had seen the judiciary as the main obstacle to the redistribution of wealth throughout society. By 1976, the 9th Schedule, initially introduced by the First Amendment to prevent undue litigation on matters of land reform, was to shield 178 articles from judicial review, almost twice as many as it had contained in 1974. Despite attempts by the Supreme Court to uphold the principles of judicial intervention, the Emergency would be used, both 'legally' and 'illegally', to intimidate and constrain high court judges and generally to weaken judicial independence on the grounds that they were 'indifferent or hostile to a socialist government'.

Yet, while the 38th and 39th amendments limited the extent of the Supreme Court's authority, they could not prevent the court striking down parliamentary amendments on the grounds that such amendments violated the spirit of the constitution. Moreover Indira Gandhi had tacitly recognised the right of the Supreme Court to judge the validity of the 38th and 39th amendments, a right it promptly exercised. Despite government attempts to amend legislation dealing with the selection and transferring of judges, the Supreme Court ruled that the executive could not direct the states to appoint district and circuit court judges against the recommendation of the high courts.

On 21 August 1975, Indira Gandhi stated in an interview with the Magazine *Blitz* that the government had no plans to undertake any fundamental constitutional change, and that ideas about reconvening a 'constituent assembly' were incorrect. In July 1976, *The Hindu* carried an article stating that 'the Emergency provided the opportunity for streamlining the judicial, electoral and legislative procedures within a Parliamentary System' (*The Hindu*, 3 July 1976) and went on to say that a move away from Anglo-Saxon principles of constitutional law should be endorsed on the grounds that they were foreign to India's traditions. Such a statement appealed widely to Hindu traditionalists within Congress, otherwise appalled at the trends towards authoritarianism, and seemed to suggest a possibility of a dialogue with other parties. On 11 August, *The National Herald* reported H. R. Gokhale's statement that the Indian constitution had borrowed too heavily from 'the western model' and was now in need of substantial amendment. Gokhale went on to say that, although the Emergency was a temporary measure, 'it would be lifted only when stability had returned, especially once the Opposition was prepared to function within the accepted framework of the constitution' (*The Hindu*, 1 May 1976). Yet it was soon to become clear that this very framework was about to be fundamentally reformed.

The proposals and suggestions that coalesced into the 42nd Amendment were varied and complex. No definitive source can be located for the recommendations made outlining constitutional changes. They had been in some senses part of the climate of debate since the early 1970s. One version, distributed anonymously to the AICC meeting in 1976, related to the arguments for a 'strong centre' and a 'committed judiciary' heard long before the Internal Emergency. Others focused on the controversial role the judiciary had played in regard to social reform and proposed a series of radical changes to the higher judiciary, abolishing its ability to review parliamentary legislation, and even proposing that India switch to a presidential system. A further incentive to utilise the general conditions of the Emergency for resolving outstanding problems occurred in November 1975, when what was in affect a government ambush on the Supreme Court to review *Kesavananda Reddy* backfired badly, with the chief justice of India dissolving the bench (Austin 1999: 332). In response, an anonymous report entitled 'A Fresh Look At Our constitution: Some Suggestions' began to circulate around Congress circles. Francine Frankel had noted that 'one of the most important purposes of the 42nd Amendment, according to the law minister, had been to reverse the effect of the Supreme Court's 1973 judgment in the Kesavananda Bharati case, that parliament could not amend the basic structure of the constitution' (Frankel 2005: 566).

These various threads were picked up by D. K. Barooah when, on the 26 February 1976, he had been authorised to set up a committee under the chairmanship of Swaran Singh to instigate a wide debate upon the subject of constitutional change. The committee consisted of A. R. Antulay, S. S. Ray, Rajni Patel, Seyid Mohammed, V. N. Gadgil, C. M. Stephen, M. P. Singh, M. P. Goswamy, V. P. Sethi and B. N. Banerjee (Mirchandani 1977: 64). Despite adverse circumstances some opposition leaders organised a National Review Committee for the institution in Bombay, but the opposition as such were never consulted over any of the proposed constitutional changes, and were seriously fettered in their ability to influence the outcome of any formal parliamentary debate. There was, in essence, a deliberate attempt to merge the political expediency of the Emergency with a genuine and not necessarily malign interest in constitutional reform, but as with the Emergency itself, the complex evolution of what would become the 42nd Amendment also shows all the pathology of the regime. Despite clearly being the source of D. K. Barooah's orders to constitute the committee, and despite no doubt being associated informally with the 'Fresh Look' document, Mrs Gandhi would feel that the committee's recommendations were too conservative and would seek to intervene herself.

The Swaran Singh committee submitted its first report in March 1976. It consisted of a series of proposals, the main one being that all parliamentary amendments to the constitution would be beyond the interpretive powers of the Supreme Court. It suggested that the directive principles of socio-economic policy set out in the constitution should be emphasised over the fundamental rights by making them judiciable. The report also recommended that further state matters relating to agriculture, education, family planning measures and the deployment of Union troops be transferred from the state to the concurrent list. A second report was submitted on 22 May followed by a third submission on 27 July. The report was then submitted to the CWC and then to the AICC, where it was formally adopted as government policy (Zaidi 1976: 19–25). Significantly, however, Swaran Singh, who spoke on most of the proposals that passed through the debate, lost on the issue of transferring agriculture from the absolute control of the state lists. To the CPI in particular, this was a major defeat that it failed to empower the central government to break up the control that dominant agrarian interests had come to exercise over state governments, not just with regard to structural reform, but to the extraction of revenues in direct taxation.

Having in effect been debated on by the AICC, the recommendations of the committee were then bypassed by the prime minister's office. Although the 42nd Amendment built on the foundations of the committee, it went further. The significance of the 42nd Amendment compared with previous congress amendments lay in its sheer size, with its emphasis on structural change. When the final document was placed before Parliament, it consisted of 2 new sections, 11 new articles and had specifically amended 37 previous articles, including all the Emergency Articles (352, 353, 356, 358 and 359). It had altered the preamble, altered the relative strengths of the Indian Union through changing the state and concurrent lists, extended the life of the legislatures to six years, changed the rules concerning parliamentary quorum, and altered the procedures for the nomination of high court judges and the extent of judicial review. It inserted an entirely new chapter entitled 'fundamental duties' and – most significantly – altered the relationship between the fundamental rights and the directive principles by making the latter judiciable. In this regard, 'the 42nd amendment attacked the very foundations of the political order established at independence' (Austin 1999: 385).

The 42nd Amendment altered Articles 74, 75, 78 and the 3rd Schedule, formally binding the president to act upon the advice of the prime minister and the council of ministers. It thus put an end to the long period of speculation about the residual powers of the presidency vis-à-vis the prime minister. In the Lok Sabha, the opposition charged that 'the net effect of the 42nd Amendment is to take away the right and powers of the president, the legislatures, the judiciary (and) the states. In each case, the power is transferred not just to the central executive but to the chief executive. That is, to the prime minister' (LSD 1976: 116).

Yet concerning the directive principles, the 42nd Amendment drastically enlarged the scope of Article 31(C) initially introduced by the 25th Amendment. Consequently the directive principles would have priority over the fundamental rights if they sought to make the ownership and control of national resources more equitable, or to ensure that the economic system did not result in the concentration of wealth.

The transfer of policy areas from the state to the concurrent lists, while falling short of including agriculture and education, substantially altered the principles of centre–state relations. The most significant in this respect was that amendment which allowed the centre to deploy forces in states without the permission of state governments. Article 257(A) gave the centre the exclusive right to deploy troops into the states 'subject to the control of the Union into any Union state or Territory'. The amendments to the state and concurrent lists increased the power of the centre in that it became permissible for the centre to take over the administration of a part of a state without having to declare President's Rule (Jhabvala 1980). Concerning the balance between the judiciary and the executive, the 42nd Amendment sought to exclude the high courts from any say in the interpretation of constitutional matters or the viability of any central law. It introduced a whole tier of tribunals that would adjudicate over works of public utility or finance. Furthermore the 42nd Amendment sought to alter the criteria upon which high court judges were appointed. Articles 124, 127, 219–221 and 224 were all amended. These amendments provided for the appointment of a distinguished jurist as a high court judge even though he may not have held a judicial office or practised as an advocate for the stipulated minimum period of ten years (Jhabvala 1980).

While the 42nd Amendment reserved the exclusive right of the Supreme Court to undertake judicial review on parliamentary amendment bills, it altered the required majority of justices required to quash such amendments as *ultra vires* of the constitution. Previous arrangements had ensured that on a bench of five a simple majority was required. The 42nd Amendment increased the number of judges from five to seven and changed the voting procedure so that only a two-thirds vote could invalidate an amendment.

The 42nd Amendment was placed before a specially convened parliamentary sitting in October 1976. Opposition MPs not in detention boycotted the proceedings. What dissension took place was not reported. The amendment was passed on 30 October 1976 by a rump Parliament of 371 MPs out of 519, sitting in a House that had technically expired a year before. The president gave his ascent to the bill on the 18 December. Earlier, on 15 November 1976, the CWC passed a resolution on the 42nd Amendment stating that 'the changes will enable Parliament to legislate against anti-national activities and remove road blocks to socio-economic legislation for development purposes' (Zaidi 1976: 159).

Significantly, or somewhat tangentially, the 42nd Amendment changed the preamble to the Indian constitution to announce it as 'Sovereign Socialist Secular Democratic Republic'. This issue had not featured in the Swaran Singh reports nor had it been a feature of the 'Fresh Look' document. The idea and pressure to insert it had derived from the prime minister, who had insisted that it was a necessary clarification of her position on communalism and the invidious position that had been taken by BJS and the RSS in opposing the government. No other amendment in this omnibus bill sought to clarify, entrench or extend the meaning of secularism, although later attempts to clarify these terms would resurface during the Janata government's debate on the 44th Amendment.

The centre had continued to intervene in state affairs after the declaration of the Emergency, removing and disciplining chief ministers and party presidents. Following the mass arrest of June 1975, the Congress Working Committee had expelled all those Congressmen linked with the Young Turks and associated with, or sympathetic to, the JP Movement. Chandrasekhar was arrested while still secretary to the Congress parliamentary party. The declaration had by March 1976 brought the Janata administration down in Gujarat. The Janata Party, internally divided over the control of various subsidiary institutions, was coming under increasing pressure from the centre on the grounds that it was not implementing the government's policies. The government was eventually defeated on the assembly floor on 12 March 1976, after which President's Rule was declared. It is, however, a revealing fact that the centre did not immediately move to dismiss the state government, so closely associated with the mass movements aimed at bringing down the prime minister. It remains a curious testimony to the contradictions and confusions at the heart of the Emergency that such normal procedures continued unabated amid talk of crisis and rebellion.

The DMK government in Tamil Nadu was also brought down through a normal legislative election, although there had been a series of complaints that certain governments were not cooperating with press censorship (Home Ministry Report 1976). The death of Kamaraj had opened up channels of reconciliation between the Congress-R and the Congress-O which, combined with a split in the ranks of the DMK, brought the Congress to power in a coalition government. Interventions were no less continuous in Congress states. Bahuguna did not long survive in Uttar Pradesh after 25 June. Unwilling to cooperate with prime ministerial diktat, and too close to New Delhi to isolate his state from political interference, Bahuguna was finally removed as chief minister of Uttar Pradesh.[1] This removal occurred under a new governor who had been appointed and 'who is hardly discreet in his utterances against Bahuguna. He has often made remarks that reveal his hostility to the chief minister' (*The Times of India*, 13 March 1975). Several southern states (such as Mysore and Kerala), Jammu and Kashmir under Sheikh Abdulla, West Bengal and Maharashtra under S. B. Chavan were able to an extent to distance themselves from New Delhi while sending bogus reports or misleading information to imply cooperation over new national policies and guidelines. In some cases specific chief ministers attempted to make private agreements with the prime minister to retain sole command of a state (interview with P. N. Haksar, 1985).[2]

By November 1976, the Emergency had been substantially consolidated and had, with the passing of the 42nd Amendment, completely suspended normal procedures of government. As early as February 1976, Somnath Chatterjee (CPM) had charged in the Lok Sabha that 'the government has made the Emergency provisions of the constitution the normal provisions by which the country is governed' (LSD 1976: 104). The central legislature had been politically neutralised, even though private bills and calling attention motions had been reinstated by November 1975. Outside the legislative assemblies the opposition had been eliminated under a now all-inclusive MISA Act.

With the passing of the 42nd Amendment, the entire political system became centralised under an executive which was in turn dominated by the prime minister working in the context of a powerful secretariat. By 1976 the secretariat had infiltrated all the central government ministries, coordinating and overlooking the drawing-up of all major policy, the management of the bureaucracy, the censoring of information, the appointment of high court judges, and the supervision of detention for political and economic offenders. The Congress no longer emphasised its party identity, but emphasised its national identity. S. S. Ray stated at the 75th session of the AICC in Chandigarh in January 1976 that 'it should be known that if the Congress disintegrated, India would disintegrate. If the Congress were destroyed today, then India would be destroyed' (Zaidi 1983: vol. 24 334). Outlining this new philosophy, the prime minister remarked that the Emergency offered a 'new way of life for our country, that through discipline, unity and hard work, not thinking of different sections or not thinking of individual rights, but only an obligation to the country' (Zaidi 1983: vol. 24 334). Certainly Brahmananda Reddy was able to claim by May 1976 that in many crucial respects the Emergency had become part of a new, disciplined way of life. Yet by late 1976 there was the growing realisation that from the national level down, the Congress command structure had disintegrated. There was also growing concern that the resulting vacuum was now being filled by the machinations of the Youth Congress and the pet projects and schemes of Sanjay, in part through the sanction of the prime minister and her senior colleagues, but in part through intimidation and default.

The increasing involvement of the Youth Congress in the government's economic policy and the organisation of the party still require further research, over 30 years after the event. But it was already evident at the time that the rise of the Youth Wing further undermined communications between the centre and the states and added to the collapse of internal workings of the government. Sanjay Gandhi was able to undermine the very concept of accountable government by working through the offices of the DMC and the lieutenant governor (interview with P. N. Haksar, New Delhi, 1985; and S. S. Ray, New Delhi, 1986) on the basis of his relationship to the prime minister and to a group of political actors and chief ministers close to her, many of whom met informally in the prime minister's house. Sanjay Gandhi had, from the evening of the 24 onwards, been close to the intrigue that grew up during the Emergency. The inspector general of police cooperated with oral orders from, or alleged to be from, the prime minister or the prime minister's son, received via the lieutenant governor. This subjugated the entire police force to the dictates of the Congress. The involvement of the Youth Congress in 'storm trooper' activities aimed at intimidating various sections of the community into cooperating with government policy is now well documented (Selbourne 1977; Shah Commission 1978). These involved three distinct projects close to the heart and interests of Sanjay: a vendetta carried out against the high courts, slum clearance and the now infamous sterilisation campaign.

The government's attempts to subjugate the high courts through constitutional amendment were also accompanied by acts of violence and intimidation carried out through the Delhi Municipal Corporation and the Delhi Development Authority, in which the Youth Congress figured. Violence proved necessary in that the established norms of legal address, especially through *habeas corpus*, had proved to be entrenched within the legal profession and also throughout middle class opinion. Subsequent interpretations of the Emergency often stress the subjugation of the judiciary to the executive without much of a fight, but in many senses the lower courts and many bar associations remained deeply hostile to the change in the balance of power between the executive and judiciary.

Lawyers' houses were demolished and tenants were evicted from property rented from the Delhi Development Authority. To punish individuals who had passed unfavourable verdicts against the government, judges were often transferred to different states at short notice. The Shah Commission later noted that what happened in the month following the declaration of the Emergency with regard to the high court judges appears merely to have been an extension of the ideas conceived on the night of 25 June (by Om Mehta and Sanjay Gandhi regarding 'closing down' the courts). The slum clearance program was even more pernicious in its remit and in the way it was carried out. Orders for such action could never be traced to the prime minister or indeed the prime minister's secretariat or even the ministries of home and works and housing. Although the minister of state for works and housing was aware of the government's slum clearance programme, ministerial responsibility for the 'beautification of New Delhi', and the removal of illegal slums from green belts, lay with Sanjay Gandhi who had no official position other than heading the Youth Congress.

The most notorious case of high-handedness occurred in August 1976 when Delhi Municipal Corporation workers, aided by bulldozers and with police support, attempted to clear a predominantly Muslim slum area from around the Turkman Gate, Old Delhi (Shah 1977; Tarlo 2003). Violence erupted and the police resorted to firing. The official death toll was 6, but other sources put it at 150 while others put it much higher. The Shah Commission stated that the Turkman Gate killings 'remained the most single act of chaos and cruelty of the Indian Emergency'. The incident was the product of an uninformed and uncoordinated executive, unable to control events even within the capital, and a state unable to manage its own personnel. H. K. L. Bhagat, minister of state for works and housing, replying to a question on slum clearance, remarked: 'It is wrong that demolitions are being made for the sake of demolitions upon unapproved colonies' (See Shah Commission 1978, Chapters XIII and XIV). Indira Gandhi denied her involvement in the projects, and later was able to rehabilitate herself by implicitly shifting the emphasis onto Sanjay.

The most notorious case of Youth Congress involvement in government concerned the sterilisation campaign. By March 1976 the government had adopted a firmer line on the family planning programme by granting fiscal incentives to people willing to undergo vasectomy, particularly government employees. The government, concentrating on the metropolitan areas of India and especially Delhi, brought pressure to bear on certain professions linked directly to state patronage (mainly teachers and the lower levels of the civil service) to provide quotas for sterilisation. This led, in the circumstances of unbridled executive power and an inadequate command structure, to the overzealous implementation of already coercive policies. The setting of targets by Delhi for most of the northern Indian states encouraged competition among Congress governments. States anxious to avoid the consequences of coercive measures sent back bogus reports stating that quotas had been fulfilled or even exceeded.

On 18 March 1976, *The Indian Express* urged in its editorial that setting targets and coercing teachers was not the correct way of undertaking family planning in the long term. Cases wherein deaths had occurred because of contamination after operation or because of actual negligence were beginning to multiply after July and August 1976. Unmarried males were sterilised, as were old men, because of the emphasis upon sheer number. Travelling vasectomy units operating throughout northern India became objects of fear when they were seen moving in to certain villages (see Selbourne 1977). Organised almost exclusively from New Delhi, the campaign was geographically restricted to northern India, especially Haryana, parts of Uttar Pradesh and Rajasthan. No policy documents have yet come to light to prove that the government was committed to a coercive sterilisation campaign or to accept responsibility for what happened. Yet despite the inability of the English-speaking press to report on excesses both in the family planning campaign and in the detention of political prisoners, rumours were widespread by late 1976, and were especially reported in the venacular press.

At the same time, such inertia and fragmentation was happening on a much more localised scale. By mid-1976 the number of agents authorised to issue MISA detention orders had become so great that it was no longer possible for the home ministry to monitor all the cases even if it had been required to (interview with Om Mehta, New Delhi, 1986). The structure of the government in Delhi, and the duplication of channels of communication to the prime minister through the prime minister's secretariat, the chief ministers and Sanjay Gandhi, undermined government coherence and communication. Krishan Chad worked at complete variance with other ministries, secure in the knowledge that he retained the trust of the prime minister. At times he even operated under the presumed direction of Sanjay Gandhi beyond the coordinating functions of the prime minister's secretariat. The Shah Commission reported in 1978 that 'records of the Home Ministry reveal that the Ministry was continually pointing out to the Delhi Municipal Corporation serious flaws in its admission of MISA detainees in New Delhi' (Shah Commission 1978: 43 11.59).

Krishan Chad went on record as saying that the home minister was working at cross purposes to the prime minister's wishes and that 'the Home Minister did not have any political sense' to any of its policies. Moreover 'the officers of the Home Department used to observe that the Lt.-Governor always had his way in these matters, and that the Ministry of Home Affairs ultimately cancelled and withdrew its instructions' (Shah Commission 1978: 43 11.61).

The Shah Commission later summarised: 'Shri Krishan Chad has also said that the prime minister handed over the running of Delhi to Shri Sanjay Gandhi. He has admitted that whenever instructions were given to him...he took them to be emanating from Sanjay Gandhi' (Shah Commission 1978: 45 11.70)

With whole areas of policy unaccountable for, factions within the Congress were increasingly concerned about the complete lack of any ministerial responsibility in areas like censorship, sterilisation and slum clearance and political detention. The CPI was also anxious to protest and to be seen to protest over excesses, but the arenas in which such acts could be carried out had become increasingly marginal. Inderjit Gupta (CPI) stated in the Lok Sabha 'I know that even the speeches of the prime minister are being censored. Who is responsible? Who is to account for this? Where is the Minister responsible?' (LSD 1975: 44)

Even if questions were raised in the Lok Sabha on matters of detention and police torture – as they were after the Monsoon session of 1975 – they could not mobilise public opinion or prompt the courts while the blanket suspension of the fundamental rights was in operation under the Emergency, and no press reported on them. Paradoxically, the greatest damage was done through the rapid circulation of rumours and the inability of the government to effectively dismiss them.

Indeed much of the overt centralisation proved counterproductive. Rumours related to excessive use of MISA flourished in conditions of censorship. As early as October 1975, the home minister issued a circular to state governments requesting them to show restraint in their use of Emergency power and undertake reviews of all MISA detainees (*The Hindu*, 24 October 1975). By October 1975, Reddy commented in Bangalore that 'Indira Gandhi was deeply concerned over reports of public harassment and abuse of the

Emergency by state officials' (*Amrita Bazar Patrika*, 7 November 1975). This warning was reiterated again in November, as the central government appeared to appreciate the opportunities the Emergency was providing for the unscrupulous use of local state institutions against the poor. By July 1976 the number of MISA detainees had risen from a pre-Emergency figure of 6,010–34,630, of which 28,386 were detained under Section 16A of the MISA Act (i.e., the grounds of their detention were not disclosed) (Ministry of Home Report 1976–1977). Concerning the Defence and Internal Security Act, 39,832 people were arrested between 25 June 1975 and 15 April 1977. Of these, 33,736 were actually prosecuted and 16,371 were later convicted. Significantly the ministry report went on to state that 'there is no systematic collection of information by the Government of India regarding action taken by the state Governments and Union Territories under the Provisions of MISA or DISA' (Home Ministry Report 1976–1977: 32).

The 42nd Amendment was an attempt to institutionalise the Emergency into the framework of the Indian constitution. Alone, it could not enable the government to create a socially viable, district-based organisation capable of implementing policy. During the statutory motion on the proclamation of the Emergency in July 1975, a member of the opposition had remarked 'it is not enough to say now that you are going to implement. What is necessary is to know what machinery you propose to set up in order that implementation is carried out?' (LSD 1976: 67).

One of the failures of the 42nd Amendment to reconstitute the Congress within the context of a permanent emergency lay in the inability of an untrammelled executive to discipline and revitalise the Congress as an organisation capable of directing popular participation in economic and political reform. By November 1976, the Emergency had stopped having any significant influence upon the economy. The return of inflationary pressures and the rise in prices generally implied that, contrary to the government's propaganda, the Emergency's relationship to the economy was largely transitory.

The wholesale price index began moving upwards in August and increased in September 1976. Wholesale price indices for agricultural goods in the first week of September were higher than the prices for the same period in 1975. *Eastern Economist* suggested that this meant 'that all the gains made in the price situation as a result of the anti-inflationary policies of 1974 and reinforced by the Emergency have been completely wiped out' (*Eastern Economist*, 24 September 1976: 604). By October, the wholesale price index was back up to 344.5. During the same period there had also been an increase in the money supply. On 6 November 1976, Romesh Thapar commented that 'to consolidate the economic gains of the Emergency may now prove a futile exercise given that prices are now gradually increasing. Moreover, given the fickle nature of the monsoons, the task of consolidation is likely to be more and not less pressing' (*Economic and Political Weekly*, 6 September 1976).

In retrospect, it is intriguing to calculate at what stage the leading political elite behind the Emergency, Mrs Gandhi, her posse of chief ministers and the critical overlap between these and Sanjay's Youth Congress realised that it was not sustainable. It is also intriguing to contemplate how they had envisaged climbing out of the hole they had so comprehensively dug for themselves. It is clear that by early 1977, key political collaborators such as the CPI were increasingly distancing themselves from the government, and especially Sanjay, and that senior Congressman were increasingly concerned over the scope and role of the Youth Congress. Indira Gandhi was conscious of both these trends (interviews with P. N. Haksar, 1985, and Om Mehta, 1986).

The factional alignments that had emerged within the Congress parliamentary party during the J. P. Narayan movement had not been dismantled during the Emergency as much as simply silenced, and of great significance later, although political parties had been banned and key leaders arrested, the state proved incompetent in silencing them completely, and in closing down the social movements and organisations that acted as broad conduits between such parties and society as a whole. Although the RSS had been banned shortly after the declaration, it was clearly organising cultural events and rallies in key parts of north India by early 1976. There was every indication that the organisations' abilities to operate under the ban had even led to improvements in the recruits of volunteers and a growth in the number of *shakhas* (Home Ministry Report 1976).

Having decided that the powers of the prime minister under the Emergency made any attempted rebellion futile, Ram and many of the 'machine' Congressmen he represented bided their time in the belief that, even after the formal adoption of the 42nd Amendment, the Emergency was but a drastic interregnum that would not fundamentally alter their interests or their positions within the party (interview with Jagjivan Ram, New Delhi, 1986). For its own part, the government had insisted that the 42nd Amendment did not substantially alter the constitution other than asserting the powers of Parliament to amend it along lines laid out within the directive principles. On several occasions Mrs Gandhi had reiterated that the government was happy with the political system, and with the general workings of federalism, and had no intention of changing to a presidential form of government, or preventing further elections once order had been restored.

Ram clearly believed in the logic behind the constitutional amendments, seeing them as a continuation of the struggle between the courts and the government over land reform started in the late 1960s. He also claimed to believe that the Emergency was an interregnum. What he could not accept was the growing strength and presence of the Youth Congress and the influence of Sanjay Gandhi (interview with Jagjivan Ram, New Delhi, 1986). There was reason to believe that, having failed to invigorate the old Congress system under her control, Mrs Gandhi was intending to use the Youth Congress as a sort of cadre-based substitute for the old party organisation. At the AICC meeting in Guwahati in November 1976, the Youth Congress dominated the proceedings and appeared as the clear political successor to the parent organisation. Moreover Sanjay Gandhi was emerging as Mrs Gandhi's clear political successor. The prime minister declared simply that 'the Youth Congress has stolen our thunder'. The Guwahati session led to a serious disagreement between the Youth Congress and the communists, with the former accusing the communists of excessive interference and calling for more economic incentives for the private sector. The Youth Congress also called upon the government to liberalise further licensing policies in the public sector and scale down planning. The opposition to public sector and state involvement in the economy directly undermined the communist's main rationality for supporting the Emergency, as did the Youth Congress' contempt for the poor.

Calls by the Youth Congress for the continual opening-up of the economy were received favourably in big business circles, keen to support Sanjay Gandhi's anti-communism as a defence against further nationalisation moves by the Congress. Indeed it is around the Youth Congress that there begins to emerge the first signs of business leaders seeking out access to the state to prevent any further radicalism, and in particular to curb the communists. There is circumstantial evidence that large industrial houses, which had funded the Youth Congress separately after 1974 (Commission of Enquiry into the Maruti Affair 1979), began to approach Sanjay directly and independently of the Congress by 1975 and 1976. Some business leaders were convinced that the Youth Congress had come of age under the Emergency and ought to be cultivated on the grounds that it would increasingly dominate the government and government economic policy.

Yet there was, by 1976, no sign of a coherent anti-communist, free market ideology emerging within the Youth Congress. Indeed there was little sign of any coherence at all. By early 1977 it was becoming evident that Sanjay's 'boys' were drawn from middle class, unemployed graduates along with criminalised elements within the informal sector of the economy, working alongside boss men like Om Mehta, Bansi Lal, R. K. Dhawan and V. C. Shukla. Through these friendships, such forces had direct access to the prime minister and the prime minister's secretariat. Most of the dubious politicians that Mrs Gandhi had adopted after 1974 were in some ways connected with Sanjay (Selbourne 1982). These people, initially occupying marginal positions within the government, flourished in the atmosphere of the Emergency.

Whatever the influence of Sanjay Gandhi on the actual declaration, conditions of arbitrary demand, duplication, minimal coordination and an all-pervading fear soon began to affect the morale of the business communities, no matter how much they had appeared to benefit from price deregulation and an atmosphere conducive to shedding excess labour (Bardhan 1984). Financial support for the Youth Congress from industry came increasingly through self-interest and fear. Interestingly, fear of Indira Gandhi and of Emergency 'socialist radicalism' soon gave way to a general concern over the behaviour of the Youth Congress towards party funding and the general defence of property from arbitrary interference by people claiming to represent the state. In such circumstances, the Emergency provided the opportunity for the use of the state, even central state institutions, as a form of 'private property' aimed at extracting resources for personal gain on an unprecedented level (Shourie 1997).

The chaos and irresponsibility of governmental decision-making under the Emergency could be dismissed only in that the centre remained in the hands of a few individuals clustered about the prime minister with no lasting control over ministerial power or without any significant political clout in the states. Yet, with the rise of the Youth Congress as a praetorian movement, the danger lay in that the otherwise arbitrary coercion of the Emergency would begin to converge towards some form of permanent authoritarian structure, linked not to any ideological project such as the abolition of poverty but simply to naked self-interest.

In 1978 the Shah Commission commented 'if this country is to be rendered safe for future generations, the people owe it to themselves to ensure that an irresponsible and unconstitutional centre of power, like the one which revolved around Sanjay Gandhi during the Emergency, is not allowed to come up ever again, in any shape or form or under any guise' (Shah Commission 1978: 119 13.311).

The rise of the Youth Congress and the decline of the CPI alliance with the Congress was an event of profound significance for the future of the Emergency. Everything about the declaration, its economic policies and its political and institutional reforms had been witness to the spontaneity behind its adoption and the necessity, once it had been called, to convert the Emergency into a charter for socio-economic reform. It was through the 20-point programme that an attempt was made to link the Emergency to the populism of the early 1970s. The Emergency had been called on the grounds that the non-CPI opposition were frustrating socio-economic reforms by working outside the constitution, and were representative of conservative and communal opinion. It was continued because, after solving the immediate problems faced by the prime minister, it appeared initially to allow the centre to move forward against long-standing problems.

As far as the prime minister claimed, the holding of elections had never been in doubt. The 42nd Amendment had made the formal continuation of an Emergency irrelevant. The vacillation and hesitancy of the government over the decision to call elections lay primarily in the deadlock between elements associated with the Youth Congress and the Congress Party proper. As early as July 1976, Reddy had stated in Madras that the economic and political objectives of the Emergency had been achieved and that elections would be held sooner rather than later (*The Statesman*, 13 March 1976). Om Mehta stated in November that elections should not be held until at least 1978 and

Sanjay Gandhi expressed his views that elections should be cancelled until the country's economy had stabilised (interview with Maneka Gandhi, New Delhi, 1986).

By December the government was seriously divided over the issue, with two clear factions attempting to influence the prime minister. Opposition came from within the Youth Congress and from the people established immediately about the prime minister. Their opposition came from the realisation that the way the government had operated under the Emergency was not suited to a system based upon popular support. The Youth Congress had little or no organisational presence at the state and district levels. Those Congressmen who realised that the Emergency was collapsing, and who feared that they might be deselected from their seats the longer the process of speculation went on, supported calls for early elections. The excesses of the family planning programmes, the slum programmes and the police generally were just the tip of the iceberg of arbitrary decision-making and petty, political vendettas damaging the state administrations in the north and isolating the south, although there was no way in which the regime could adequately gauge the scale of the problem or access the likely impact it would have. These issues and themes would provide the basis for the massive political mobilisation that would take place against the Congress.

A switch back to an electoral system would strengthen those elements within the Congress Party who had been paying lip service to the Emergency and who would be able to regain their political clout through the selection of candidates. It has been suggested that there is some evidence that Mrs Gandhi called the elections to allow the Youth Congress access to the government and the Congress parliamentary party, establishing what had become the 'typical' basis for Sanjay to secure the prime ministership at some future date (interview with Haksar, 1985). Sahgal has convincingly suggested that this idea could have been based upon Mrs Gandhi's own experience in 1969 wherein it was the staunch support of the parliamentary party that ensured her victory against Desai and ultimately against the syndicate (Sahgal 1982: 179). Certainly this conjecture is supported by the dramatic inclusion of Youth Congress candidates for the 1977 elections.

On 18 January 1977, Indira Gandhi announced to the nation that the Lok Sabha was to be dissolved and fresh elections were to be called. It is still unclear in what forum her decision was debated, if at all. Many of her senior colleagues expressed surprise. Certain provisions of the Emergency were to be relaxed and members of the opposition still in detention were to be released. It is quite reasonable to assume that the decision was a personal decision of the prime minister taken in the belief that the Congress could still win, and that the time allowed for the opposition to organise a national campaign was inadequate to ensure victory.

D. K. Barooah was called to the prime minister's house on 16 January and informed that elections were to be held (interview with D. K. Barooah, New Delhi, 1985). Barooah pointed out to the prime minister the fact that the party was not in a position to fight a general election and that it could not be sure of victory. The prime minister had replied that she was in possession of Intelligence Bureau reports that gave the Congress at least 220–270 seats (interview with S. S. Ray, 1986). Indira Gandhi had also pointed out that the opposition was still badly divided and would not be expected to present a credible challenge at the centre. She considered the position of the national opposition as similar to that in 1971, disregarding the experiences of Gujarat, and the importance the RSS/BJP in providing the ground work through which to effectively mobilise a campaign at short notice. The prime minister then discussed her decision with the cabinet on 17 January.

The Congress Party set about organising itself for elections in late January. The process of allocations for candidates was undertaken with the Youth Congress demanding a large number of the tickets. The Pradesh Congress committees had been authorised to forward their lists of prospective candidates to the centre by no later than 28 January with the Central Election Committee scheduled to meet on 2 February. The resulting allotment of seats under central nomination to newcomers largely from the Youth Congress caused panic amid the established leaders. Rumours circulated that the Youth Congress had been awarded up to 150–200 of the 542 Lok Sabha seats, and that the Bihar list had been seriously amended to the final exclusion of Jagjivan Ram's supporters (interview with Jagjivan Ram, 1986). Large allocations to Youth Congress members throughout Uttar Pradesh led to the resignation of former chief minister, Bahuguna, from the Congress. In turn, Bahuguna played an important part in pressing Jagjivan Ram into resigning. He was successful only when he managed to show Jagjivan Ram the list of candidates for Bihar drawn up by the Congress Parliamentary Board under the prime minister (interview with Bahuguna, New Delhi, 1986). This precipitated Ram's dramatic resignation from the government on 27 January, followed by 35 Congress MPs, to form the Congress For Democracy (CFD).

These two events alone indicated the mood in the Indian Hindi belt; they also revealed the extent to which an attempt, however contrived, to realign state–society relations through the medium of elections presented a critical space for former collaborators of the regime to defect at such a critical moment. Ram's decision and subsequent criticism of the Emergency caused charges of opportunism and consternation from the government but it fundamentally transformed the situation confronting the government. Announcing his decision, Ram issued the following statement to the Press: 'the internal democracy of the Congress organisation at all levels had not only been abridged but had almost been abolished...most dangerous procedures have been adopted to topple some chief ministers who disagreed with the dictates of some individuals' (*The Statesman*, 3 February 1977).

The CWC that met on 2 February to discuss the issues raised by Ram's resignation noted that 'he was directly associated with the decisions of the Congress. For him now to repudiate all his responsibilities and repeat the allegations of the opposition parties against the Emergency does not resound to his credibility' (Zaidi 1983: vol. 24 237).

Ram's resignation gave remarkable impetus to the non-communist opposition parties in their attempts to form a national version of the 1975 Janata Party in Gujarat. He controlled significant factions in Uttar Pradesh and in Bihar, and his willingness to undertake seat adjustments transformed the electoral credibility of the opposition. For Congress, his resignation led to an immediate reassessment of candidate lists and a sidestepping of the Youth Congress. The formation of the Congress For Democracy broke the spell that somehow the Congress was invincible by revealing that the Emergency had never commanded universal support within the party in the first place. Most significantly, as a dalit, Jagjivan Ram had been of great symbolic importance to the Congress and to the prime minister. As the court case had provided the one issue for the non-CPI challenge in the first place, so the excesses of the Emergency became the one political issue that drew the Janata alliance successfully into the 1977 campaign behind J. P. Narayan's ideological and philosophical leadership. Again, as in the Gujarat campaigns, the RSS and the BJP stepped into the breach to assist the parties coming together, their societal presence having been completely transformed by the long and public opposition to the Emergency, and by the extent to which their volunteers had suffered from detention.

Following the announcement of elections, MISA and censorship laws were no longer enforced (although Gokhale reminded several newspapers about the voluntary agreement on ethics). The home ministry reasserted its control over the law and order machine by ordering the majority of political detainees to be released, despite protests from the prime minister's secretariat. After 18 January, the alternative 'ministry' of the DDA was restored to its purely municipal functions as ministerial responsibility was reasserted and several key officials were replaced. By the time of the polls, the number of MISA detentions had been reduced from 34,360 to 10,903. Three days after the announcement of elections, the government withdrew the powers devolved to the states under Rules 11, 14, 31A, 43, 69 and 71, wherein police powers could be used without prior central approval. But MISA remained until the election result indicated a substantial Congress defeat and was revoked by the prime minister just before she offered her resignation to the president.

It will never be known what would have happened had the Congress won or, more interestingly, had the Janata/CFD coalition secured a wafer-thin majority amid charges of widespread rigging. It is reasonable to assume that in the case of a Congress victory the Emergency would have remained, although it is questionable how much it could have achieved or what it could possibly hope to achieve on the basis of the 1975–1978 experience alone. In the event of a hung parliament it seems probable that an attempt would have been made to cling to power through some variant of the method used in 1975. Yet the result was too unambiguous. It was impossible in March 1977 for the Congress – or the followers of Mrs Gandhi – to have ignored the political judgement of the electorate except through the deployment of massive coercion and through a coordinated response by key institutions of the state such as the police and even the army. Although that force was available, it was unlikely that the social groups and interests within the party and throughout India that had implicitly accepted the Emergency of June 1975 would have supported such a move a second time around. The 1977 elections were perceived as a transformative moment in India democracy. Many years after these events, Partha Chatterjee noted that:

> What was distinctive in the life of Indian democracy was, I think, the defeat of Indira Gandhi's emergency regime in a parliamentary election. It brought about a decisive shift in all subsequent discussions about the essence and appearance of democracy...the 1977 elections established in the arenas of popular mobilisations the capacity of the vote and of representative bodies of government to give voice to popular demands of a kind that had never before been allowed to disturb the order and tranquillity of the proverbial corridors of power.
>
> (Chatterjee 2004: 49)

The experiences of the Emergency revealed starkly the limits placed on the centre's ability constructively to intervene within societal affairs, and even to regulate the coordination of specific institutions within central government. Romesh Thapar remarked soon after the announcing of elections that authoritarian rule had proved to be more difficult than had been first imagined. The Emergency as an experiment in centralised rule had led to increasing divergence and chaos. As Rajni Kothari commented in *Seminar* in April 1977 'it is just not possible to rule over this diverse, far-flung and multi-layered society; to implement policies, even to maintain 'order and democracy' by concentrating power and decision-making in New Delhi; much less in a small secretariat' (Kothari 1977).

The sixth national elections dealt the Congress its first all-India defeat. On a national turnout of 60.5 per cent, the Congress' popular vote dropped to 34.5 per cent, with the Janata Party, consisting of the Congress-O, the BJS, the HLD and the Socialist Party, capturing 41.4 per cent. Congress failed to take any Lok Sabha seats in Bihar, Haryana, Himachal Pradesh, Uttar Pradesh, and Delhi and Chandigarh Union Territories. In Rajasthan it managed to secure one seat. The 'Janata Wave' was repeated later in 1977 and in 1978, when a Janata-nominated Union president – Sanjiva Reddy – the man Mrs Gandhi had sought to prevent becoming president in 1969 – dissolved 17 assemblies throughout India. In those states that had experienced political instability in the late 1960s and the worst 'excesses' of the Emergency, the Janata Party took control.

6 Political mobilisations 1980–1986

What had struck so many commentators of the Emergency at the time was the speed and comparative ease in which democratic governance was collapsed into authoritarianism (Dhar 2000). And yet such ease was testimony to the creeping nature of authoritarianism in India since the early 1970s, the fact that this degree of centralisation was made possible by the constitution itself, and a certain evident anti-democratic streak within middle class, elite opinion. Citing Arun Shourie, Granville Austin notes that '[a]lthough the emergency, in the extensiveness of its evils, was an aberration in the history of Indian democracy, it was also "the culmination of long tendencies". The centralisation of authority grew from the constitution and the central-command structure of the Congress Party' (Austin 1999: 297).

Political leaders had centralised the state in order to deliver an emancipatory, socialist agenda to society as a whole, but it had been continued as a strategy to protect a specific leadership and to keep a government in power. In the context of the Congress split, and the ongoing political mobilisation at the regional level, the centralisation became incoherent, pragmatic rather than strategic. Authoritarianism became a device to shore up Congress votes. The need to redefine the local and regional levels of governance was ignored for the short-term expedient of providing a national mandate for the Congress-R. In the context of the 1975 Allahabad court case and the specific challenge to the prime minister, this contingency of using state institutions to protect the office of the prime minister was carried out ruthlessly, culminating in the declaration of the Emergency and the first series of constitutional amendments carried out by an emasculated parliament.

The degree of chaos at the heart of the Emergency was to a large extent disguised by the effort taken to choreograph the event through constitutional niceties and procedures. In comparison with other third world or post-colonial states, where the degree of preparation and execution of the seizure of state power has tended to be much more overtly coercive or crude, this coup was constitutional. Such elements were present in the immediate run-up to the declaration of the Emergency (the threat to close down the courts, cutting off power to the press), but what is striking is the effort that went into legitimating the declaration, and the blizzard of ordinances and legislation that sought to facilitate and justify the extension of executive power. Given the degree of ongoing political mobilisation, and the relative success with which the ideas and practices of democracy had taken root in Indian society, the Emergency regime had to represent itself as a necessary requirement of state policy, with the JP Movement made to represent a threat of disorder, a 'mobocracy'. The position of the RSS and the BJS within the JP Movement was of particular use to the Congress-R, since it could be represented as a form of Indian fascism, organised throughout society in a series of seemingly innocuous cultural and social organisations, biding its time for the opportunity to strike at the Nehruvian state itself, and remove his daughter's government.

Mrs Gandhi personally favoured this view (Frank 2001: 217) (interview with P. N. Haksar, 1986). The CPI were supportive of her interpretation, and bought into the government's ideological reading of a progressive 'socialist' government versus communal, right wing 'reactionaries' linked to the old traditionalists in the Congress-O. Having grasped the nettle on the evening of 24 June, the prime minister reiterated the legitimating themes of garibi hatao but faced the same dilemma she had confronted after 1971: how to reform the state, and having reformed the state, how to co-opt societal interests that had been mobilised by the ongoing breakdown in the Congress system, and in particular by Mrs Gandhi in 1971. It was within this impasse that the Emergency finally came to rest. Frankel concluded at the end of her work on India's political economy that 'on the whole, the Emergency was perhaps less remarkable for official excesses against individuals...than for the limitations on government ability to use such absolute powers in bringing about basic social change' (Frankel 2005: 566).

There were to remain, around the prime minister, political elites that proved reluctant to, even arguably incapable of, abandoning a specific historically determined state structure, and anxious at every turn to ensure some essential element of constitutional legitimacy and continuity with India's reformist past. Yet others saw the Emergency, no matter how contingent and defined by the crisis of the prime minister and the ruling party, as an opportunity to push for radical redefinitions in state–society relations and in democratic practice itself as a consistent policy of advancing naked self-interest. These new emergent ideas of India and its necessary policy direction were to become part of a conflict within the Emergency regime, and within the institutions of the Indian state, and they became more concentrated as the central state became more isolated from lower state institutions, and from significant sections of society. It is in this context that Sanjay Gandhi, the Youth Congress and the challenge to the CPI were to become such pivotal events.

As the Emergency progressed, it became bizarrely ideologically inconsistent – indeed almost non-ideological. Having achieved its immediate ends, the new regime had no coherent vision of what it should then do, or how it could go about doing it. Although willing to interfere in aspects of the state such as the judiciary, the bureaucracy and the police, Indira Gandhi's vision of an India in which the state was able to carry out a fundamental overhaul of social inequalities was not different in outline from Nehru's. Although novel in terms of political style, Indira Gandhi's nationalisation strategies were not out of keeping with the earlier policies that had established India's mixed economy in the 1940s. As the 1975–1976 period was to amply demonstrate, despite the enormous powers that the Indian state acquired under the Emergency, it did not move to further alienate large business houses or to extract coercively resources from the rich farmers in order to fundamentally realign societal power. Ironically, as the coercive powers of the regime tightened, the people who suffered were the very constituencies for which the Emergency had been deemed necessary: the urban and rural poor, the scheduled castes and tribes, and the minorities (Tarlo 2003).

It is the sheer degree of state centrism that explains the sort of Bonapartist language that dominated the first wave of literature aimed at analysing the Emergency. What interests did the Emergency represent other than those of Indira Gandhi's need to save herself? Did the moves towards some form of corporatism, seen dramatically in union reform and arbitration, reveal a bias for big business? (Rudolph and Rudolph 1987: 159). Was it a middle class coup? Was it a coup in defence of the newly established farmers and cultivating castes? Was it an attempt to include the lower levels of the caste and jati systems and break the inherent conservatism of change by a reformist parliamentary democracy? In a sense it was all these things, at different times, and in different places within the state–society nexus, as the political elite around Mrs Gandhi sough to re-engage and establish some form of collaborative strategy.

As the Emergency began to unravel, the prime minister decided to call for elections. The results were unprecedented. Commenting on these a decade later, the Rudolph's noted 'in 1977, Janata's victories in both parliamentary and state assembly elections seemed to mark the end of the one party dominant system. Congress's thirty year hegemony over India's national government and federal system was broken when Janata governments took office in Delhi, and in nine of seventeen large states' (Rudolph and Rudolph 1987: 159)

The electoral defeat of the Congress was absolute in areas that had been most effectively penetrated by the central government, and most exposed to both the sterilisation campaign and excessive detentions under MISA. The Congress Party's survival in the south indicated that in these states it had been able to distance itself from the worst influence of arbitrary central government. It had retained the ability to implement state policies and was able to retain popular support. It was this lack of local credibility, underlined by casual coercion and brutality in the north, that led to defeat as the Congress lost the votes of scheduled castes and tribes, and Muslims in particular. Dominant peasant groups switched to a variety of parties, mostly Charan

Singh's BKD – renamed the BLD, although some voted for the BJS faction of the newly formed Janata party. A resolution by the CWC in April 1977 somewhat understated the reasons for the extent of the Congress collapse in the Hindi heartlands: 'It has been recognised that things happened during the Emergency in a manner which alienated the people from the Congress organisation and the Government' (Zaidi 1983: vol. 24 265).

The Working Committee went on to discuss the defeat in Bihar with particular reference to a failure to enact land reforms and to the fact that the organisational machine had been 'defective'. Congress lost all 139 parliamentary seats in Uttar Pradesh and Bihar. In both Orissa and Punjab non-implementation of the party programme was held to be the main reason behind the electoral defeat. The failure of the district committees over northern India, especially in Uttar Pradesh, Punjab, Bihar and Haryana was the direct result of the dismantling of the state Pradesh committees through central intervention and their replacement by nominees from 1969 onwards. Little or no mention was made over excesses by the Emergency regime until 1978 and the widespread debates stimulated in part by the provisional findings of the Shah Commission. When Brahamananda Reddy was elected president of the Congress Party in May 1977, he remarked: 'as a result of the long spell in power, the Congress organisation at the centre and the states tended to lean too much on the government apparatus and, as a consequence, to neglect the traditions of mass involvement' (Zaidi 1983: vol. 24 227).

The electoral victory of the Janata reversed this trend dramatically: it mobilised into the central state politicised but differentiated interests that had been mobilised by frustration and coercion and a decade of Congress hyperbole. Such interests were aggregated into one party at the national level but one that still operated at the regional and local level as separate political entities, increasingly sectional in interest. Many of these factions and splinters belonged to political formations that had begun to manifest themselves in the late 1960s.

Indeed, in broad terms, the election of the Janata Party was a return to the trends indicated by the 1967 elections, but accelerated by the experience and drama of the Emergency itself (Rudolph and Rudolph 1987). The parties that came together to form the Janata Party were constituent parts of the Grand Alliance that had taken on Congress and had then, in 1971, sided with the Congress-O (again represented by Desai and a powerful collection of Hindu traditionalists). The decision for a full-out merger – as opposed to pre-electoral seat adjustments or parliamentary support – seemed a decisive break with the past, however, creating the semblance of a two-party system of government in India for the first time. The decision of the CPI and the Left in general to cooperate with the newly formed Janata Party significantly maximised their chances of victory.

Forged by the mutual experiences of imprisonment by Mrs Gandhi, the Janata set out to create a viable ideological alternative to Congress centralism, premised on a neo-Gandhian manifesto that stressed the importance of agricultural and local government institutions (especially with reference to rural institutions), appropriate sustainable technologies and food for work programs that stressed the social uplift dimensions of JP. In this sense it sought to correct an urban bias, and did represent a powerful kulak class who asserted their position with the newly formed party and the government. Such policies were also in keeping with Congress traditionalists that sought to combine socialism with a 'saintly idiom' of politics linked to both J. P. himself and, more generally, to Desai (Morris-Jones 1967). Such a program created a common framework within which the former political parties could work to restore formal democracy, but more importantly seek to accommodate the widespread surge in popular expectation in the wake of Mrs Gandhi's defeat.

Significantly, the involvement (and momentary emersion) of the BJS in the Janata 'experiment' did not involve any emphasis on Hindu nationalist themes favoured by the RSS and the BJS in the 1960s. The articulation of a socio-economic agenda by the former BJS, led purposefully and moderately by Atul Behari Vajpayee, drew on the party's newfound legitimacy as a democratic movement, and articulated some of the populism that had been part of the 1971 campaign, especially with reference to peasant proprietors and middle traders. For a party long associated with fascism, the kudos acquired by the role of the RSS volunteers in resisting the Emergency is not to be underestimated. The emphasis on a neo-Gandhianism, drawn from the Congress (and in particular from Congress traditionalists like Desai himself), appealed to the leaders of the party as well as to sections of the RSS, and they especially approved of Desai's candidature for prime minister (Rudolph and Rudolph 1987: 169).

Yet the fusing of the BJS, the BLD, the Socialist Parties (led by people such as Fernandes and Narain), the Congress For Democracy and several dissidents from the Emergency regime itself (most notably Chandra Shekhar and Mohan Dharia) distracted attention from the fragmented nature of the party and its regionalisation, if not so much in terms of ideology, then in terms of the ruthless ambition and competition of the leadership and, by implication, their 'faction' of the party. Each former party's share in the overall Janata vote is instructive – the largest share went to the former BJS faction, with 90 seats in the Lok Sabha, and 31 per cent of the national vote, the Lok Dal was second with 55 seats and 19 per cent of the vote, close to the socialists who obtained 51 seats and a 17 per cent share of the national vote. While ostensibly unified at the centre, many of these parties were competing in the same states for the same votes (Hewitt 1989).

The electoral dominance of the BJP was not reflected in its share of ministerial places or its position within the organisational structures of the Janata. Yet because of its large share of the Janata vote, the BJS attitude to merging was crucial. It had been its refusal to even condone coalition government in the late 1960s that had been blamed for the collapse of coalition governments in the states because of their isolated, doctrinal view of politics, and their hostility to other parties. The apparent self-effacing quality of the former BJS at the national level was to provide them with further kudos as mediators of problems between leading personalities of the Janata, as they emerged from 1977 onwards. Yet the RSS retained its separate institutional Sangathan structure, and as a consequence of this, many of the other parties – especially the socialists – refused to merge their own social and youth movements into one formal organisation. Despite the formation of an overall Janata party structure, with Chandra Shekhar as president, there was little cohesion to what was, in many ways, a much weaker version of the Congress Party's earlier eclecticism, but without any of the historical weight of a nationalist party structure able to pressurise for compromises and results.

Lacking a clear and dominant leadership, the Janata also proved incapable of using the institutions of state that Mrs Gandhi had used, increasingly after 1971, to realign factional squabbles and to discipline state governments. It was ironic that, having roundly condemned the expanded prime minister's secretariat, Desai retained its functions – and yet even its implicit, 'extra-constitutional centralism' did not provide him with the ability to coordinate his government. It was to be of great significance at the time that Desai became prime minister without a formal vote by the Janata parliamentary party. Whatever programmatic unity did exist at the centre, personal animus would provide the context in which struggles for power framed wider ideological disputes at the regional and local levels. In this sense, the literature on the Janata phase is right to stress the politics of 'vendetta and revenge', but as the lines of conflict are traced from Delhi down into the states, the disputes became more class based and increasingly more between the former Jan Sanghists and the former socialists.

The most intense dispute took place between the former BJS elements and the former BLD under Charan Singh, as the two parties sought to consolidate their position within the same caste alliances of cultivating peasants – the so-called kulak (Corbridge and Harriss 2000 81–3). Charan Singh and Vajpayee in effect brokered a deal that divided up the newly elected state governments between them. Former BJS chief ministers emerged in Madhya Pradesh (MP), Rajasthan, Himachal Pradesh (HP) and New Delhi, with former Lok Dal leaders heading up Janata administrations in Bihar, Uttar Pradesh (UP), Haryana and Orissa (Rudolph and Rudolph 1987: 168). The election campaigns were poorly coordinated, with factions relying on their separate youth and student wings, and failing to coordinate expenditure through the Janata party organisation. Disputes were mediated not through institutional procedures, but through personal intervention on an ad hoc basis.

The electoral ferment within the north Indian states made these elections particularly violent, with a dramatic increase in caste violence, especially involving upper castes and harijans or dalits, and a rise in Hindu–Muslim rioting (Home Ministry Report 1978–9: 122–31). Such turbulence was symptomatic of the intense competition that the Emergency had incubated and revealed the extent to which dominant cultivating caste 'farmers movements' set out to capture state resources and to exclude others, utilising street politics and direct mobilisations (Chatterjee and Mallik 1997: 67–73; Corbridge and Harriss 2000). In areas such as UP and Bihar, where backwards castes were sufficiently numerous and organised to resist forward caste politics, the switch to post-Emergency, post-Congress politics was also marked by intra-caste violence between castes of similar economic status, but of differing ascriptive ranking (Brass and Vanaik 2002). The result was a sort of 'amoral castism', in which collective action for the greater good – and despite the wider emancipatory language of 'saving India for democracy' – degenerated into what Austin later called a 'survival society' in which individual castes move to defend their interests 'indifferent to the well-being of others' (Austin 1999: 641). Such a process, disguised by the large majorities of 1980 and 1984, was to mark the beginnings of an intense and sustained political and cultural fragmentation that would characterise the 1990s (Mitra 2006).

Social violence would provide the context in which ideological differences within the Janata Party over secularism would emerge to dominate national politics, as well as provide the basis for the restoration of Congress credibility under Mrs Gandhi. One of the most dramatic cases of caste violence occurred in July 1977, in the remote Bihari village of Belchi. Here dominant agrarian castes massacred a large number of dalits who had been encroaching on the ritualised status of caste Hindus by not showing due deference and by seeking to extend their rights. The national and state government's response proved to be inadequate, and given the isolation of the region, it took some time for the scale of the disaster to emerge (Frank 2001: 419). For the newly installed national government, it was unfortunate that the first politician to arrive at the scene, in darkness and after fording a flooded river on the back of an elephant, was Mrs Gandhi. This dramatic political act, highly fortuitous if not directly politically staged, began a slow reconciliation between Mrs Gandhi and what had been a key constituent of her 1971 victory – dalits. This one apparent act of courage brought about a curious rapprochement between Indira Gandhi and J. P. Narayan, which, followed up by an emotional and symbolic visit to her former constituency, marked the beginning of Mrs Gandhi's electoral comeback (*The Times of India*, 12 August 1977).

Indira Gandhi's political rebirth, and the indication that she was seeking to mobilise support so soon after the Emergency itself, acted as a solvent on the factionalising structure of the Janata Party. She was to depict the government as vindictive to her and her family as well as politically incompetent, captured by powerful agrarian interests who did not care for the poor and who were contemptuous of their well-being. Amid incidents of Hindu–Muslim violence, she consistently accused the communal elements within the Janata – the former BJS, and the still-existing relationship between them and the RSS – as being primarily responsible. She consistently sought to question the government's secular credentials. Both Congress and the CPI effectively played on public fears, and those within sections of the Janata Party, that the neo-Gandhianism of Desai was providing a cover for the widespread infiltration of 'communalists' into the state, and facilitating a 'secret' RSS plan for India.

The Janata Party did achieve significant results in amending – indeed restoring – forms of democratic governance, but its ideological eclecticism meant that this was often inconsistent and compromised by the new complexities that defined national politics. Mrs Gandhi herself had revoked the Internal Emergency before she resigned as prime minister. Janata revoked the External Emergency of 1971, repealed legislation against censorship, restored the principles of electing numerous local and municipal authorities and set up a series of important government committees on matters as diverse as caste reservation (the Mandal Commission) and panchayat reform, all of which were to prove influential. The Janata party also – mostly through the efforts of its socialist elements, and the active support of Charan Singh – set up a Minorities Commission as a bulwark of secularism. Attempts to repeal preventive detention proved less successful, especially in the context of a deteriorating law and order situation across north India. Although having revoked the External Emergency, and having allowed preventive detention legislation to lapse while in government, Charan Singh was to reinstate it under the guise of the National Security Act in December 1980 during his brief tenure as caretaker prime minister after the Janata government had fallen.

In other areas too, the new government sought to restore the status quo ante. Despite initial wrangles over senior judicial appointments (especially that of chief justice of India, following Chief Justice Beg's retirement in February 1978), the Janata government restored the principle of seniority, and countermanded the transfer of high court judges that Mrs Gandhi had implemented as a way of punishing signs of judicial independence. With reference to federalism, despite a clear commitment to the states (and the successful revoking of the centre's powers to deploy force within states without their consent), the Janata government started its short life by setting the unfortunate precedent of dismissing all the Congress state governments and calling for elections. It did this on the basis that the unusual circumstances of the Emergency had not so much discredited Mrs Gandhi's central government, but through her centralised rule, Congress administration in the states.

Several Congress state governments applied to the Supreme Court to countermand the decision to dismiss them on the basis of being *ultra vires* of the constitution and that a 'gratuitous' use of Article 356 undermined federalism. The court ruled in favour of dissolution, setting an unfortunate precedent that would be used against the state Janata governments in the wake of Mrs Gandhi's national victory in 1980. Only in the mid-1990s, in radically changed circumstances, would the Supreme Court extend judicial review to validate the reasons given by the centre for any proposed declaration of President's Rule (see below).

The 42nd Amendment was successfully amended through the omnibus 44th Amendment, which was introduced into Parliament by Janata in May 1978. Having rejected a simple annulment of the 42nd Amendment by a single piece of legislation, the 44th Amendment was mauled by cross-party talks between Janata and the Congress, still dominant in the Rajya Sabha. As such Congress was able to retain some elements of the 42nd Amendment, namely the privileging of the directive principles over the fundamental rights by retaining Article 31(C), and the retaining of tribunals to deal with specific acts of judicial review (Austin 1999: 427). Thus Article 31(C) as amended by the 42nd Amendment was retained, preventing the Supreme Court from regulating and restricting Parliament's powers of constitutional amendment until the Supreme Court struck it down in 1980 in the Minerva Mills case, thus returning India to the status quo ante of the Brahamanda Reddy 'basic structure' doctrine (Austin 1999: 427). The 44th Amendment had managed however to tighten up procedural issues relating to presidential discretion vis-à-vis the cabinet, and had removed the ambiguities surrounding the Article 352(ii) definition of an internal emergency.

Yet aside from restoring the credibility of the constitution, and re-establishing a relatively clear separation of powers between the differing branches of government, the government was dominated by how to deal with Mrs Gandhi, and how – in the face of an aggressive Congress comeback, to hold up the factional collapse of the party. In this respect, rising to the challenge as effectively as she had in 1969 and 1975, Mrs Gandhi's political skills proved especially formidable in the absence of any basic coordination within the government, between various central institutions, and the ambition of individual politicians. Even the Shah Commission, which began hearings in September 1977, and which managed to effectively record the excesses of the Emergency, provided a forum for Mrs Gandhi to manipulate the media. Her initial refusal to attend, her eventual reluctant appearance on the stand, as much as the spectacle of her being arrested by the police in October, created powerful iconic images that transformed her public position into one of victim and even martyr (Dhar 2000, Frank 2001). Charan Singh's determination to punish her and to pressurise the Delhi police through the home ministry led to the first glaring failure of collective cabinet responsibility within the new government and resulted in her release following inadequate charges and insufficient thought to the legal process (Frank 2001). Punishing a woman for being a dictator by disregarding due process struck the press as being both ironic as well as tragic (The Hindu, 12 September 1977). Charan Singh's failure to respect the prime minister's decision over how to deal with Mrs Gandhi led to the first serious breach within the top leadership of the Janata.

Having split her own party again in opposition (to form the Congress-I) in 1978, Mrs Gandhi was returned to the national parliament in 1978, in a by-election from the Chikmangalur constituency in the southern state of Karnataka. She was briefly prevented from taking her seat in the Lok Sabha following a ruling that she had previously (during the Emergency) abused parliamentary privileges. She was eventually allowed to participate in the proceedings. The government collapse began even as the 44th Amendment was being introduced, when Charan Singh sought to undermine Morarji Desai as prime minister, seeking to use the home ministry (and beyond, his political base in Uttar Pradesh) to publicise allegations of corruption against the prime minister's son. Other disputes broke out between the socialists and Chandra Shekhar referring to the allocation of candidates in state elections, and between the former Lok Dal supporters and former Jan Sanghists.

In April 1978, Charan Singh resigned from the party organisation, a move which signalled factional strife within the BLD-led governments of Haryana and Uttar Pradesh against supporters of Desai, but also against former BJS leaders. This dispute, given the balance between the BJS and the BLD within Janata as a whole, was to be much more significant, although it is not immediately clear that it was ideological. (Austin 1999; Frankel 2005). The rapport between Desai and the Hindu Nationalists presented them with an opportunity to increase their influence within the government as a whole, and certainly gave them an incentive to support Desai against Charan Singh, but it threatened chaos in the states, and was to generate concern among the former socialists over the role of the RSS and the future of secularism in India, principally Raj Narain, a close associate of Charan Singh.

As Charan Singh and Desai argued over the merits of bringing Mrs Gandhi to justice (as well as Sanjay and the main posse of leaders around Sanjay), serious rivalry broke out at the state level between the former BJS and the BLD. This in effect broke the deal struck in 1977 over the control of chief ministerships. Charan Singh resigned from the government on 30 June 1978, ostensibly over the issue of arresting Mrs Gandhi without cabinet approval. He was later readmitted into the government in January 1979, but this was only a temporary stay until the government finally collapsed in July. The final break between Charan Singh and Desai came in the context in which former BJS state leaders brought down three former BLD chief ministers and staked their claim as a faction of an increasingly moribund national party. In retaliation, and significantly, invoked by Raj Narain, Charan Singh played on the issue of secularism, and sought to depict Desai as working with 'communal' forces – a reference to the RSS – and revived the controversy of the so-called dual member issue – demanding that members of the Janata party should resign from the RSS. Here he made common cause with Mrs Gandhi herself, who continued to argue that the RSS was 'undemocratic' and 'secretive'.

By the beginnings of the 1979 Monsoon session, Morarji Desai faced a no-confidence motion in government, and resigned as prime minister on 15 July 1979. In a letter to the president, he requested permission to form another government. Citing convention (drawn in the main from a British model of a two-party system), President Reddy called the leader of the opposition Y. B. Chavan, leader of the rump of the former Congress-R following Mrs Gandhi's creation of Congres-I, to try and form an alternative administration. After failing to produce evidence of a clear majority, the president ignored the counterclaims of Jagjivan Ram, and invited Charan Singh to form a government who had, in the interim, secured the offer of support from Mrs Gandhi's Congress-I. At the critical moment – when it came to test his majority on the floor of the parliament – Congress-I declined to support him.

It was widely known at the time that the conditions for support were the abolition of the Shah Commission and the dropping of all charges pending against Mrs Gandhi and her son, Sanjay. Despite being in effect a caretaker prime minister, and despite having failed to test his government before a formal vote of the Lok Sabha, Singh advised the president to dissolve parliament and call fresh elections. In a moment of almost operatic irony, President Reddy, whose selection as Congress presidential candidate in 1969 had provoked Mrs Gandhi's decision to split the party, had created the context in which Mrs Gandhi would return to power.[1]

The Janata Party collapsed ostensibly through the bitterness and ambitions of its leading politicians, but there was a wider cause to the instability that such rivalry ignited, especially at the regional and local levels of north India. Jaffrelot is right to note that although secularism and the RSS was central to the collapse of the Janata Party, the ideological dispute was heavily manipulated by central politicians – especially Charan Singh. As the Rudolphs had stated earlier, 'while it is not clear that the secularism issue exhausted Charan Singh's and Raj Narain's motives, it was the issue which they used to carry on the power struggle against Desai and, in time, to justify the successive splits that destroyed the party' (Rudolph and Rudolph, 1982: 141). In government, Atal Bihari Vajpayee had emerged as a credible foreign minister, flexible and sensitive to the Pakistanis, and to many, a moderate and capable politician. It is difficult to read the collapse of the Janata as a straightforward fight between secularists and communalists, but again it is necessary to place the contestation in the context of a complex state–society model that provides a sort of multi-causal, non-linear explanation of events.

The myriad state structures were galvanised by the sheer degree of electoral instability that was the hallmark of post-Emergency India, provided the RSS and the former BJS with an opportunity to capitalise on their new found legitimacy. The defeat of the Congress party in 1977 significantly weakened the central state's ability to impose secularism as a hegemonic ideology, in Jaffrelot's terms, since it brought to power a government that contained not just Hindu nationalists but a significant number of Hindu traditionalists around Morarji Desai. Desai's definition – and support – for the RSS as a cultural organisation reflected Patel's sentiments in the late 1940s, and provided a niche in which the large number of shakars – which had appeared to have expanded and strengthened during the Emergency to around 13,000 (Jaffrelot 1996: 288) – could operate with impunity. The growing synergy between Hindu traditionalists and nationalists was evident in a series of policy areas that were in the main overshadowed by the infighting within the government.

Amid parliamentary debates to clarify the meaning of 'secularism and socialism' as inserted into the constitution by the 42nd Amendment, the Desai government showed itself particularly sensitive to former Jan Sanghists. Having decided to retain the changes made my Mrs Gandhi to the preamble, elements within the government wanted secularism defined to exclude Hindu nationalism by identifying it as in effect synonymous with communal ideology. Desai brokered a deal that provided a sufficiently narrow definition of communalism, which would enable former Jan Sanghists to vote for it and still be able to articulate *Hindutva* as a key signifier of India's cultural identity. In the end, the adopted definition was defeated. Attempts to move cow slaughter from the state to the concurrent list of the constitution (thus facilitating a national ban), found favour with Desai, and his own personal interest in banning alcohol found favour with the Jan Sanghists and the RSS, as did his interest in Ayuvedic medicine and the much ridiculed urine therapy practiced by the prime minister (Park 1976: 521–22).

Desai supported a private members bill, introduced by the ex-Jan Sang MP O. P. Tyagi, seeking to prevent religious conversions based on 'force, fraud' or other material inducements, an issue that would return to dominate India in the early 1980s and again in the mid-1990s. Although all of these specific initiatives failed, they revealed a shift in what can broadly be called the 'cultural policies' of the state under Janata that created a space in which long-cherished policies of Hindu nationalists could seemingly be set out and debated without immediate censure. Prefiguring some of the most significant controversies of the late 1990s, Desai also proved willing to side with former BJS politicians and a separate campaign by the RSS calling for the removal of a number of controversial textbooks from high school and university curricula on the grounds that they gave a misleading and distorting view of Indian culture and history (Adeney and Saez, 2005; Rudolph and Rudolph 1982: 131–54).

The controversy started in mid-1977, when Desai's PPS sent a memorandum to the ministry of education drawing attention to a series of academic texts written by eminent historians (two texts involved Romila Thapar), in which aspects of medieval India had been written with a lack of sensitivity, or the 'requisite fervour' with regard to religious aspects. The analysis of the event is complicated by the prime minister's dislike of what they believed were leftist, Marxist-inspired historiographies that debated the 'Muslim period' in terms of material class exploitation and not through its religious identity. The controversy weakened the secular credentials and furthered the image of the BJS as a Trojan horse within the government, linked to the RSS that, in being removed from the political arena, was unaccountable to the electorate while being in a position of growing influence (The Times of India, 6 August 1977).

Such convergence between Desai and an emergent ethno-religious agenda – whatever the nuances between them – provided critics of the prime minister, and indeed critics of the government, with a stick with which to beat the secular drum. It marks the slow, insidious process where the language of secularism displaced the debates on socialism and poverty. By late 1978, increasing communal tension across north India and the apparent involvement of RSS volunteers in certain riots, specifically Aligarh and Jamshedpur (Engineer 1985; Home Ministry Report 1979), provided Mrs Gandhi with an opportunity to identify the RSS (and in the same debate, the recent textbook issue) as creating a climate of violence against the minorities which would become the norm. Again, this played into the divisions opening up within the government. It led to growing tensions within the national Janata Party, with Chandra Shekhar and Madhu Limaye seeking to distance themselves from the former Jan Sanghists, with his eventual formation of the Janata-Secular.

Mrs Gandhi's condemnation of the RSS was, even before her return to power, shot through with the ambiguity of a personal style that now seemed more accommodating of Hind religious values, which in the wake of her meeting with J. P., now regularly involved visiting temple sites and other public endorsements of devotional aspects of Hindu religious culture. Her successful portrayal of the Janata government as one incapable of preventing societal violence also referred to intra-caste, intra-Hindu violence, as much as the violence between Muslims.

However paradoxical, the beginnings of a Hindu votary first appeared during Mrs Gandhi's final term in office. There is considerable evidence suggesting that it comes into force through ongoing societal instability, namely a change in the economic and social status of minorities, and a concomitant growth in the sense of Hindu vulnerability to both lower caste and backward caste forms of political assertion, as much as that by non-Hindus (Hansen 1999; Varshney 2002). Specific symbolic crises – mainly issues of mass conversion – would conflate these concerns wherein members of the backward and scheduled castes would mobilise against the stigma of their low ritualised status by becoming Buddhists, Christians or Muslims. In differing ways, the RSS, the newly formed BJP and the victorious Congress-I utilised a growing mood of anxiety and crisis to further their specific agenda, and in doing so contributed to the deepening of such anxieties and legitimised their presence within the language of the state itself. In Chatterjee's earlier phrase, the state elite appeared to confront a specific configuration of societal forces that for differing and often contradictory reasons appeared to favour religiosity and the dominance of religion as a political signifier (Chatterjee 2004; Hinnells and King 2006).

In 1979–1980, this shift was by no means self-evident. On the face of it, the experience of the Janata government had strengthened the RSS dislike for politicians, and their views on the divisive nature of party politics aimed at capturing the state through tactical populist socio-economic strategies (Jaffrelot 1996). Thus the rebirth of the Bharitya Janata Party in 1980, which lay claim to the traditions of the BJS but strengthened and widened through the Janata experience, inherited an ambiguous, somewhat aloof relationship with the RSS. Deoras returned the RSS to an emphasis on a more voluntary, activist strategy and concentrated on extending the shakhas and creating non-political fronts that would directly mobilise Hindu sentiments without necessarily antagonising the new Congress-I government. The BJP's adoption of Gandhian socialism owed a great deal to the influence of the Janata period, and although linked to Upadhyaya's 'integral humanism' of the late 1960s, seemed to mark a divergence between the new party and its RSS base.

The new party's refusal to use the word Hindu in its founding documents contrasted sharply with the move of the RSS to overtly religious organisations such as the VHP, which Deoras effectively relaunched in 1979, charged with the specific job of creating a Hindu votary through religious reform and the public mobilisation of Hindu symbols (Jaffrelot 1996; Vanaik 1997). The proliferation of new forms of non-governmental organisations that bloomed in the wake of the Emergency (Chatterjee 2004) provided a conducive background to diversify what would later become widely known as the Sangh Parivar.

Founded in 1964 amid calls for a 'world Hindu Conference', the VHP emerged as a principle forum through which to standardise Hindu religious practices and beliefs; to 'brahmanise' as well as to standardise sources of Hindu religious activities in keeping with other organised world religions, a process of emulation that lay at the heart of Hindu reform from the late nineteenth century onwards (Jaffrelot 1996; Jones 1989). Amid the apparent lack of concern within the BJP for key issues of religious nationalism, the RSS actively promoted the societal sangathanist strategies of the VHP, and by 1984 its new student wing, the Bajrang Dal. Together, these organisations – within the overall Sangh Parivar – would increasingly articulate Hindu religious concerns through overt ethno-religious mobilisation through youth movements, educational foundations and union and student activities, with scant attention or debate over social and economic policy. The VHP as a social movement – not the BJP as a political party – would lie at the heart of the construction of a palpable sense of Hindu loss, and it would require the BJP to reverse its post-Janata emphasis on socialism in order to benefit from the VHP's societal mobilisation by the late 1980s. The context for such a rapprochement would be engineered by the Congress-I and its eventual slow death as the single dominant party through the next decade.

The rise of *Hindutva* starts, in the first instance, through the electoral instability and divergence that characterised India after 1977. Seeking to link such divergent – and diverging – theatres of political activity to a national party structure became increasingly tenuous and volatile. It called for the open acceptance of national coalition government, and for the recognition that states would often increasingly be of a differing political hue to the government in New Delhi. This was, of course, the very model of Indian federalism to which Mrs Gandhi was opposed in 1969 and again in 1975. In 1980, the circumstances of her return to power implied to her that she could still succeed in reversing the trends exemplified by the Janata Party, despite the overwhelming evidence of the 1970s (Haksar 1979).

The restoration of Congress-R in 1980, and the victorious return of Mrs Gandhi as prime minister, was largely expected, the only surprise lay in the extent of the Congress-I majority. The 1980 elections confirmed superficially what had become known for a time as the 'wave' pattern of Indian electoral support (principally 1971, 1977, 1980 and 1984) when, in comparison with the indecisive 1967 elections, the polls produced categorical results and large working majorities. In 1971 Congress obtained 43.7 per cent of the popular vote, in 1977 it dropped to the lowest level of 34.5 per cent and returned in 1980 to 42.7 per cent (Butler 1985). In 1984, in the highly ununusual circumstances of Mrs Gandhi's assassination, Congress achieved a record 401 seats, over 70 per cent of the Lok Sabha, on something approaching 49 per cent of votes cast. Yet there is evidence that these results were, in a sense, anomalies, triggered by national events, and were made possible by the fluid and transitory nature of a majority of India's political parties, a view confirmed by the national elections of 1989 onwards. Although dramatic, they were not restorative of the status quo ante. The Congress system was dead, as much as the highly stratified vote banks that had enabled it to function. National events initially proved pivotal for starting this electoral wave-like phenomenon, but increasingly, the regional forums, with their specific spatial distribution of caste and jati coalition, came to prevail (Hewitt 1989).

In the wake of a highly successful national election, and the decision to dismiss opposition state governments en mass (replicating Desai's decision against Congress-R state governments in 1977), the Congress-I 'wave' continued, but with considerable variation. While on paper it appeared to some academics that Congress had restored itself (Weiner 1980), closer scrutiny revealed the extent of the breakdown within the Hindi heartland, especially with regard to the long-term relationship of the Congress to the scheduled caste votes (Rudolph and Rudolph 1987; Wood 1984b). Mrs Gandhi's ability to mobilise these votes in 1971 had been instrumental to her victory. Yet in 1980, in combination with the dissident Congress-U vote, the non-communist opposition vote was greater in Uttar Pradesh than the Congress-I: 46 per cent as opposed to 38 per cent (Butler 1983). Congress won the state through the old disparity between seats and votes in multiparty contests for single seat constituencies under a first-past-the-post system. It was not lost on analysts that the continuing mobilisation of caste and class votes were moving towards differing parties, and made electoral outcomes much more uncertain. Congress success in the states after 1980 rested as much on splintering opposition parties. Scheduled castes abandoned Congress-R en masse, returning in 1980 but then defecting to non-Congress parties in subsequent state elections almost as soon as they were due (Manor 1983).

In Bihar, the non-communist opposition vote was 40 per cent compared to the Congress vote of 34 per cent. Again, even without the dissident Congress-U vote, the Congress-I secured a mere 1.4 per cent lead over the sum total of the constituent parties of the previous Janata alliance. In both Bihar and Uttar Pradesh, scheduled caste votes and Muslim votes were important, and the volatility of their support for Congress, as well as Congress overall share of the vote, revealed future vulnerability to future opposition seat adjustments, or mergers – as would happen in 1989 (Manor 1980). By 1983, no less than five major states were to be run by regional parties, with the Punjab as well voting for a non-Congress government by 1980. In 1984, scheduled caste votes swung behind Rajiv Gandhi but less through any specific and sustained affinity to policies and more as an instinctive move to the Congress in the wake of Mrs Gandhi's death (Frankel 2005; Jaffrelot 1987).

Inheriting a clarified and restored constitutional structure, Mrs Gandhi, who had actually voted for the 44th Amendment, as returned to the Lok Sabha by the upper house, confronted a party organisation that had split again in 1978, and which was in many senses merely an extension of her own political charisma. The party had ceased to exist in any meaningful sense. As with 1969, the 1978 split generated the pragmatic need to reconstitute Pradesh committees through defection and nomination, with compliant chief ministers trusted by the prime minister and, in effect, reporting to her alone. In-between the holding of elections, the party structure almost ceased to exist at all (Kohli 1990). Furthermore, the legacy of the emergency was visible in anxieties and concerns expressed by the legal profession and the collapse of several court cases still pending with reference to the Emergency, many of which, in effect, dissolved by themselves in order to avoid offending the new government. Those who had, during the Janata Party interregnum, been particularly outspoken against her and Sanjay were anxious to avoid Mrs Gandhi's vengeance. Mrs Gandhi herself undertook the recalling of the Shah Commission reports wherever this proved possible.

Soon after 1980, it became obvious that, after her victory, Mrs Gandhi remained committed to the view that the only way to run India politically was to have Congress running the centre and the states under the same leadership, with the prime minister effectively managing party factions in a style that were almost identical to that which occurred between 1971 and

1975. To achieve even the rudiments of organisational efficiency required the continuation of the interventionist centre, either through her office or more formally through President's Rule. Between 1980 and 1986, President's Rule was used on 23 occasions. In most cases, Mrs Gandhi, or those who acted on her behalf, having tried to engineer defections earlier, now sought to remove opposition parties through the governor. In the context of highly mobilised and unstable societal alliances, intervention often inspired or provoked regional oppositions and actually helped consolidate their hold on political power (Kothari 1985). In an extraordinary case in Andhra Pradesh, the dismissal in 1983 of the Telegu Desam chief minister, N. T. Rama Rao, while he was receiving medical treatment in the US was so ham-fisted that he was later reinstated following his presentation of his legislative majority to the Union president in Delhi. Such crassness had been provoked in part by a series of opposition conclaves and meetings calling for widespread constitutional change and involving predominantly southern states under the auspices of Ramakrishna Hedge, chief minister of Kanartaka, as well as Rama Rao and later Farooq Abdullah of Jammu and Kashmir (Austin 1999: 541–43).

This hesitant regrouping of the non-communist opposition, excluding for a time the BJP, marked a return to the broad trends that had characterised the Indian party system since 1967, and again after 1977. A significant difference between these earlier experiences and what Austin calls the 'constitutional revolt' of the early 1980s was an emphasis on coalition and power sharing (as opposed to merger à la Janata) as well as a sustained criticism over the constitution and a call for change, especially with reference to the role of the governor and modifications to President's Rule. Dismissed by Congress-I as irresponsible, Mrs Gandhi nonetheless responded in 1983 by setting up the Sarkaria Commission to report on centre–state relations and – after undertaking the widest consultations possible – to make a series of recommendations. It was a task that failed, both to elicit real enthusiasm from the prime minister or indeed to convince the opposition that the central government was prepared (or indeed capable) of working with non-Congress political formations (see Hewitt 1989; Manor 1991). Mrs Gandhi's final period in office was never free of allegations of authoritarian design. Such allegations took on either a rumour or threat of another form of constitutional crisis such as the Emergency, or an attempt to undertake constitutional reform and install a presidential system of government (Adeney and Saez 2005; Austin 1999: 495)

On their own, such constitutional musings would have been harmless, but in the context of specific 'toppling' operations carried out against state governments, and amid renewed controversy over issues such as judicial independence, the transfer of high court judges and debates on preventive detention and freedom of speech issues, they struck the opposition as both ominous and familiar.

In the wake of Sanjay Gandhi's death in July 1980 following an air crash over South Delhi, Mrs Gandhi appeared to become more indecisive and isolated (Franks 2001: 448–50). Indicative of the drift that set in at the centre was the government's failure to renew any of its main socialist manifesto commitments made in the 1980 general election. The period from 1982 onwards is marked instead by a gradual, hesitant abandonment towards private, market-led strategies that would provide the basis for Rajiv Gandhi's period in office (Ahluwalia 1995; Balasubramanyam 1984: 202–04; Mooji 2005). Indira Gandhi retained much of the Janata's emphasis on small-scale industry and paid particular attention to agriculture, but there was little attempt at structural reform. There was no sense in which the new government sought to re-engage with the commitments of the early 1970s. One analyst noted that 'Indira Gandhi's last term in office as prime minister was virtually devoid of major economic policy initiatives' (Rubens 1985).

The consolidation of the farmers lobby – dramatically illustrated during the final days of the Janata Party by Charan Singh – prevented any further reform in agriculture, and through agitational politics and Gandhian techniques of 'resistance' the lobby pressured governments to increase subsidies and to support prices which did not by themselves assist the poor. Excluded from tax revenues, Congress agricultural policies took the line of least resistance between dominant social interests and the poor, whose position had in some cases actually deteriorated during the 1980–1984 period (Corbridge and Harriss 2000; see Lipton 1995: 327–40). Industrial and urban interests, irritated and alarmed by bullock cart demonstrations regularly bringing metropolitan India to a standstill, responded by creating a miasma of anxiety and concern over public order (Chatterjee and Mallik 1997: 80–100; Corbridge and Harriss 2000).

This failure by the national government to respond to social exclusion and the rise of unstable – often populist – governments in the states (see below) strengthened the growth of both the voluntary sector, especially with reference to welfare programs and education, and the provision of anti-poverty programmes (Heston 1990). State governments followed pre-electoral largess, often involving subsidies to key social constituencies as vote winners, which increased both the public debt and led in many cases to unsustainable borrowing from the centre (Joshi and Little 1993: 31–52; Kohli 1988). Many non-communist party structures also mirrored the 'populist-authoritarian' leadership of the Congress, dominated by a single leader, and often with pretensions of dynastic succession (Kohli 1990). India's dramatic growth in NGO activity, which while to some extent indicative of organised social interests better equipped to deliver emancipatory projects, represented an abandonment of state policy to small-scale and often unaccountable organisations that often brought about further state decline (Kohli 2004; Kothari 1993). As such, a new emphasis on NGO activities, with reference to education and social welfare, flourished (Kohli 2004; Lipton 1995). As local state institutions fell further into decay, or became more blatantly rent seeking in their behaviour, successful voluntary organisations, particularly in municipal areas, paradoxically brought about further state rollback and urban differentiation, with the poorest areas least able to organise and thus reverse a deteriorating and neglected urban infrastructure (Basu 2005).

Yet perhaps the most telling legacy of Mrs Gandhi's final term in office was her apparent switch from a determined defence of secularism to an equivocal support for communal politics. Jaffrelot cites this selective accommodation – characterising her policies towards Punjab, Assam and other states of the North East – as clearly undermining the secular norm that had dominated national governments from the time of Nehru onwards. While this strategy was made possible by the contradictions implicit within the Nehruvian definitions of multiculturalism and modernisation, between individual rights and group rights derived from religious and cultural categories, it was above all a sign of political weakness amid the collapse of the central state's commitment to emancipatory polices. These contradictions were contingent to the shift in strategy followed by Mrs Gandhi: they did not determine it. In mid-1983, the veteran political commentator, Romesh Thapar, noted that 'the vibrations of Indira Gandhi's recent posturings in Assam, Punjab and Jammu...have sparked speculation that the Congress-I is to be fashioned into a tribute of Hindu assertion'. (Thapar cited in Jaffrelot 1996: 314)

In the Punjab Mrs Gandhi's attempt to restore Congress government involved a tactical compromise with Sikh extremism and a fatal estrangement with Sikh moderates such as Darbara Singh, who accused her of seeking to generate anxiety within Punjabi Hindus for electoral gain (Frankel 2004: 677). Political agitation in the Punjab was not new, and the issues of Sikhism's relationship to Hinduism, and Punjab's contribution to and share in federal 'wealth', were long-standing issues. Yet by the late 1970s there had been, in keeping with the general constitutional agitations noted above, a sustained move to clarify Sikhism as a separate faith, and to resolve outstanding issues left over from the creation of Haryana in 1966 with reference to a shared capital and shared water resources (Tully and Jacob 1985: 36–51).

Mrs Gandhi's response to this complex state-based agitation was either to manipulate issues of ethno-religious identity to appease allies or to prevaricate while others did so, in order to destroy a non-Congress political party and restore her own party (and faction) to power (Grewal 1990: 205–27; Tully and Jacob 1985). Ignoring the advice of the Congress chief minister of Punjab, who urged the arrest of Sant Bhindranwale, Mrs Gandhi sought the advice of Zail Singh (former chief minister of the Punjab 1971–1977 and Union president 1983–1988), who throughout remained 'more concerned with undercutting Darbara Singh to maintain his influence in the Punjab, than in cooperating with the chief minister to eliminate a potential threat to Indian unity' (Frankel 2005: 671). The consequences for this – Indira Gandhi's repeated failure to arrest and charge Sant Bhindranwale before he took sanctuary in the Akal Takht in the Golden Temple – were literally catastrophic. It was seen by many as a tacit use of communal violence to polarise moderate Sikhs and Punjabi-speaking Hindus to vote for the Congress. By October 1983, Sikh extremists were killing Hindus and moderate Akalis and the state was under President's Rule (Frankel 2005; Singh 1991: 357–90; Tully and Jacob 1985: 52–71).

Shorn of the specific details of the Punjab, this tactic of intervention, prevarication and then action was by now quintessentially Mrs Gandhi's style. The symbolic language used made little reference to socialism or emancipation but contained an implicit reference about minorities and their threat to the integrity of the nation. That such a threat was in itself the product of central interventions around the logic of mobilising votes was well theorised by a number of academics (Brass 1994; Kohli 2001). Frankel notes with reference to the Punjab that 'Mrs Gandhi's failure to act sooner does not easily fit into an explanation of moving to a Hindu communal appeal' (Frankel 200: 673) yet the unintended consequences of her move, and the appalling consequences of her death, enabled Sikhs to be stigmatised as a threat both to the Hindus and to Indian unity.

Again, given the implicit contradictions within Indian secularism and given the ambiguities between religious and socio-economic rights, how and in what ways had these communal idioms become so salient? It has been argued here that one of the main contributory factors was the electoral mobilisation of the poor by the state, and the resulting decline of the central state's ability to manage, direct and sustain such a process free of crisis. By the mid- to late 1970s, overtly authoritarian populism – the peak of which constituted the Emergency – had undermined the central state's ability to police and prescribe certain forms of political activity even as it moved to insert the term 'secular' into the constitution. At the same time, the Emergency generated societal forces situated around the RSS and the Hindu nationalists able to resist the state and demand recognition. Yet this can only be, by itself, a partial explanation. Tropes of Hindu loss and the sentiments of Hindu majoritarianism had been present within the Congress since the time of Patel, but they had not been determining. So who or what released the furies of Hindu communalism (Vanaik 1990, 1997)?

At some stage – and this is the key to the Congress miscalculation – Mrs Gandhi (and the extent to which she used her principle advisors) must have been sufficiently concerned about Hindu nationalism to feel it necessary to effectively co-opt or neutralise it (Vanaik 1990). On what basis did the newly elected prime minister arrive at this conclusion, if not by 1980, then certainly by late 1983? Religious issues and anxieties had, by themselves, surfaced within society by the time the new Congress government was installed, dramatic enough to suggest opportunist strategies for retaining political power – especially in the states.

The most dramatic incident involved the mass conversion of Muslims in southern India in early 1981. Here 1,000 scheduled caste members of the village of Meenakshipuram converted to Islam, culminating in a ceremony that laid the foundation for a new mosque in the village itself. Coming in the wake of the (failed) Janata Party discussions on banning forced conversions, and the protracted debates over defining secularism during the passage of the 44th Amendment, these events caused a sensation. Further political drama followed when it was alleged that large sums of money – most from sources outside India – had been given to Muslim proselytising organisations, amounting to what one national English language paper hailed as an 'International Islamic Conspiracy for Mass Conversion of Harijans' (cited in Jaffrelot 1996: 341). Several articles in the press alleged that the president of the Jamaat-I-Islami was present at the foundation-laying ceremony itself, and identified Arab states as being implicated in fund-raising (Banerjee 1982).

The event, one of many that was to follow throughout 1980s in quite rapid succession, reinforced the idea of a Hindu majority under threat, but also one in need of urgent internal reform and reintegration. The RSS response was to draw attention to the need to improve the socio-economic conditions of harijans within Hinduism as a religion (an implicit criticism of failed Congress policies aimed at eliminating poverty) in order to remove the economic and social inequalities that made harijans susceptible to financial inducements. The RSS response was also to redouble its efforts to integrate 'practising' Hindus into a reformed national culture, explicitly drawing on an ecclesiastical structure capable of defending the 'faith and the nation'. In turn, the VHP, not the BJP, provided a vehicle through which sadhus, concerned over the 'threats' to Hinduism, could mobilise such issues unmediated by existing party-based elites and structures. The VHP and its youth wing, the Bajrang Dal, emerged here as the key organisation within the Sangh Parivar by 1984. The shocks that radiated out from Meenkashipuram arguably revived similar debates and issues that had dominated Hindu politics in the 1920s: that Hinduism was too weak and to divided to stand up against aggressive foreign faiths, and that through conversion and polygamy, Muslims would become the majority population, as early as 2231 AD (Jaffrelot 1996: 342).

Public opinion polls, which were becoming increasingly sophisticated and ubiquitous in the press, showed that there was high public interest, especially throughout urban north India, to legislate and prevent such conversions – 78 per cent in one specific poll (Jaffrelot 1996: 342). The response by the Congress government was to instruct the home ministry to increase surveillance on specific Muslim organisations, and to seek to bring forward legislation aimed at preventing 'forced' conversions – especially where financial reward was offered as an inducement – but without seeking to compromise individual rights.[2] Through the anger and concerns generated by Meenakshipuram, Mrs Gandhi consciously sought to limit the gains that the RSS and the VHP in particular might make from such an explosive incident. As such, the conversions were just one incident that drove Mrs Gandhi towards what was a largely contingent rendezvous with *Hindutva*. As Jaffrelot documents so convincingly, a whole series of additional events played on Hindu concerns regarding vulnerability; from Meenakshipuram, the Akali Dal agitations, the civil disorder and violence that dominated north India until as late as 1988 (the highest yearly figure for Hindu deaths in Punjab), and related problems of 'illegal' Muslim immigration in parts of the north-east as well (Kohli 1990; Varshney 2002).[3]

Yet the most significant event was the nature of Mrs Gandhi's own assassination, the resulting communal frenzy that took place in Delhi from October to November 1984 and the emergence of Rajiv Gandhi, heading up a massive Hindu majority, having throughout the campaign explicitly invoked the image of Hindu majoritarianism (Vanaik 1997; Varshney 2002). In the wake of Mrs Gandhi's death, the law and order situation in New Delhi effectively collapsed for a period of three days. Estimates vary widely over the actual number of Sikhs killed during the riots – with the most conservative putting them at just over 1,000, and the highest at 3,000. (There were to be ten subsequent commissions and committees of investigation[4]). In keeping with earlier historical riots on this scale, and significantly, to anticipate the degree of communal rioting that would dominate India from 1989 to 1993 and again in Gohara, in Gujarat, in 2002, clear evidence existed as to the complicity of the local police, as well as the role of local party officials – in this case – Congress-I, in instigating and directing the violence (Varshney 2000, 2002). In such a highly charged environment, the Congress-I stood to gain as the polarisation of communities yielded significant – if not short-term – electoral gains. Amid the articulation of outrage and revulsion at the horror caused by the rioters, there emerged a clearly discernable language of punishment and revenge on a particular community for threatening to undermine the territorial, economic and cultural fabric of India (Robinson 1997; Anand 2007.)

The virtual eclipse of the BJP in the 1984–1985 elections is also instructive. Having started out deliberately to represent socio-economic reform and socialism, Atul Bihari Vapayee was confronted – and outmanoeuvred – by the significant shift of Hindu concerns behind the Congress. The nature of Rajiv's campaign, playing on the fears of the disintegrating nation, was one that attracted implicit RSS support, and there is evidence that the organisation actively encouraged support for, and campaigned on behalf of, Congress-I (Jaffrelot 1996; Vanaik 1997).

The election campaign as a whole deepened the base of political mobilisation further. In 1984, electoral participation in rural areas overtook urban turnouts, confirming that Indian democracy was one in which 'the turnout of the lower orders of society is well above that of most privileged groups' (Frankel 2005: 626). Yet having been returned with a four-fifths majority of seats, a number that exceeded Nehru's majority in 1952, the

Rajiv government did not seek to systematically shift to the Hindu right, to in effect occupy a space made possible by the societal reactions to Mrs Gandhi's murder. Nor did it seek to reinforce any lasting link between the radical socialist heritage of Mrs Gandhi and the reformist appeals of socio-economic redistribution. Once in power, Rajiv Gandhi would consolidate an approach to communities that, while ideologically contradictory, 'had the general effect of establishing communal idioms [within] political discourse at the expense of secular themes' (Jaffrelot 1996: 322; Mitra 1995).

Between 1984 and 1989, the Gandhi government presided over a contradictory and selective use of 'protecting' religious communities but in a manner that was incoherent and purely reactive. With regards to socialism, the government reiterated the major themes but in the context of a tentative neo-liberal reform that sought to strengthen private initiative (Adams 1990). The culmination of this incremental pragmatism was to facilitate the RSS-VHP and, by 1989, the BJP's ability to run an increasingly ethno-religious campaign premised around the Babri Masjid Mosque at Ayodhya. Ultimately, for reasons of electoral expediency and because of its own extraordinary gaff over the Shah Bano case (see below), the Rajiv government refused to confront the Hindu nationalists directly, or to expose the socio-economic paucity of the movement's policies, but rather pursued a strategy of selective appeasement – thus retaining, and often amplifying – the impression that the dominant societal cleavages were those between religious communities (Nugent 1990). Such strategies further distorted not so much the principle of secularism but arguably the specific Nehruvian version of secularism improvised since independence (Bhagava 1997; Khilnani 1997).

Having been placed into the public spotlight by the death of his brother Sanjay in 1980, Rajiv's claim to the prime ministership lay in his family name but also, somewhat perversely, in his lack of experience within politics. Rajiv's experience in business was used consciously to construct the image of a moderniser, a technocrat who would bring to the government and the party a fresh approach, untainted by the Emergency and by previous Congress infighting (Chatterjee and Mallik 1997; Nugent 1990). Confronted immediately by the complete lack of a party organisation, Rajiv used the opportunity presented by the centenary celebrations of the founding of the Indian National Congress in 1985 to chastise the party and to call for institutional renewal. He spoke very much in the language of his mother, with the curious but revealing omission of the word 'socialism' (Chatterjee and Mallik 1997: 107–09; Mehta 1994).

Despite setting out with enthusiasm, the reform process stagnated rapidly as the prime minister confronted the sheer degree of corruption and bogus membership within the party and the absence of local structures at the district and block level. By 1986–1987, Rajiv was, through necessity, returning to his mother's conviction that the party organisation could be administered centrally, enhanced by technological innovations that would allow better record keeping and greater coordination of policy – especially through identifying the important caste and class affiliations dominant within each state, and seeking to match candidates for specific constituencies. Technology became a substitute for organised political participation and for the absence of any sustained popular involvement in deciding Congress policies. Through this technique, Rajiv presided over the further and sustained depoliticisation of the state amid rapid socio-economic and cultural change, and the deliberate neglect of any sustained developmental public policy (Adams 1990; Kothari 1995).

Significant and positive contributions were made to reinstitutionalise elements of both the political process and the political system, but on the whole they failed to re-engage the state in such a way as to direct and channel societal pressures towards any coherent program. The government was committed to amending the constitution to facilitate the creation of a series of panchayat raj institutions below the state level, and to provide them with adequate political and economic importance to encourage active (and elected) support for development projects. Although legislation was not to reach fruition until the early 1990s, it was the first attempt by any national government to initiate some form of devolution since the agricultural development programs of the late 1950s.

In the passing of the 52nd Amendment (an anti-defection bill), Rajiv sought to prevent the fragmentation and splintering of the non-communist regional parties and to prevent the rampant corruption that followed attempts to form governments in the states through the use of public largesse. Yet despite these policies, and a series of accords signed with regional parties aimed at recognising the positive role regionally based, non-Congress parties could play in government, Rajiv Gandhi could not sustain or deliver on these promises. The reasons for the sudden and extraordinary collapse of the massive mandate was in part political style but overwhelmingly an inability of the state to follow through central initiatives, or as equally significantly, correct central initiatives in the light of specific local and regional circumstances.

Rajiv Gandhi's approach to the Punjab Accord was undermined by any real awareness of the issues raised by India's complex federal system, especially the complexity of Chandigarh's shared status as capital between Haryana and Punjab, and the complex issues of shared water resources that the Accord either misunderstood or ignored. Had the home ministry been involved in the process, many of these errors – which proved fatal – might well have been avoided (Singh 1991: 391–408). By 1985–1986, Rajiv had returned to a confrontational role with opposition state governments, showing in the specific case of Jammu and Kashmir that he was as capable of engineering the collapse of a chief minister as his mother had been (Hewitt 1995).

At the centre, frequent cabinet changes weakened ministerial briefs and strengthened the tendency and influence of individual trusted 'advisors' who Rajiv approached on single issues and came increasingly to rely upon. Such a tendency, underscored by family connections in the case of Arun Nehru, opened up the channels of informal communication and brokerage that paralleled formal institutions or bypassed them completely (Chatterjee and Mallik 1997: 156–99; Nugent 1990). Such arrangements would provide the conduits for personal interventions and informal 'pressures' to influence policy outcomes. By 1989, following the revelation of so-called kickbacks to the prime minister's office from the Swedish armaments manufacturer Bofors to 'facilitate' the purchase of a new field piece for the army, the moral authority of the government swiftly collapsed (Nugent 1990). In seeking to legislate for a right to 'intercept' and monitor private postings, and in seeking to muzzle the press through an ill fated anti-defamation bill, the government appeared both paranoid and ineffective, despite its massive majority (Hewitt 1994).

With reference to communalism, the most profound mistake the young prime minister made was not with reference to the Hindus but with reference to the Muslims. The undertaking of the Rajiv government to pass parliamentary legislation exempting Muslims from Section 125 of the Code of Criminal Procedure was to generate a wave of support for the BJP that would arguably sustain it until about 1992–1993 (Adeney and Saez 2005; Mehta 1994: 93–100). Such legislation implied to the Hindus that the government had abandoned any commitment to Article 44 of the directive principles and a uniform code for a uniform citizenship, despite their earlier 'sacrifices' to the Hindu Reform bills of the 1950s.

The government's initiative to intervene in this complex legal case was, from the start, based on a naive understanding of electoral politics and the apparent importance of the Muslim vote in a series of forthcoming state elections (Frankel 2005; Nugent 1990; Rudolph and Rudolph 1987). What was at issue was the extent to which, given the failure to enact a uniform civil code as stated by Article 44 of the constitution, the state should seek to balance variations in personal law as demanded by specific religious communities against individual rights demanding equality of treatment before the law. This was an old and familiar problem in India since the time of the Hindu Reform bills, and one that successive governments had left alone, initially under the Nerhuvian misapprehension that the problem would disappear over time as members of the minorities would eventually initiate their own reform.

In 1978, Shah Bano, an Indian citizen and a Muslim woman, sought – and obtained – the right to divorce and receive maintenance from her husband under Section 125 of the Code of Criminal Procedure. In 1980, following a review of the amount paid to her (which the court increased), her husband appealed to the Supreme Court on the grounds that Shariat law should govern the issue of his original divorce and the duration and amount of alimony.

In 1985, the Supreme Court ruled that Section 125 applied to all citizens of India regardless of their religion, noting with regret the failure of all governments since 1952 to pass a much needed uniform civil code, and then added the proviso that their ruling was in keeping with the Qu'ran, and Shariat, since in their view, Islamic law provided for the payment of maintenance (Larson 1995: 258). The interpretation of Shariat law by a secular body, and the subsequent intervention into personal law by what was seen as the widening of the scope of Section 125, culminated in what has generally been identified as the largest agitation launched by Muslims since independence (Jaffrelot 2006: 335) – allegedly over 300,000 in Bombay. In responding to such demonstrations, the government ignored Muslim moderate opinion and several organised women's movements, all of which were deeply concerned over the conservative and patriarchal aspects of Shariat law, and ignored various divisions within the Muslims over sects and types of Islamic practice. Instead of reaffirming a commitment to the spirit of Article 44, the government sought an accommodation with 'a Muslim community' whose leaders it uncritically took to be conservative ulema and religious clerics, and to pass legislation that entrenched personal law further.[5]

In February 1986 the government introduced The Muslim Women (Protection of Rights On Divorce) bill, that exempted Muslim men from the need to pay alimony after a specific term – *iddat* – wherein the divorced wife would be provided for by her own family and, if appropriate, the *waqf,* or local mosque management committee. In the context of the Meenakshipuram conversions and the anxieties of Hindus over external threats to their own religious identity, the government confronted a sustained Hindu backlash in which Hindu religious leaders were able to articulate long-established grievances, culminating in the allegation that the Congress was prepared to appease Muslims at the expense of the majority community. The Hindu nationalist press argued that such preferential treatment was based on the electoral significance of the Muslims to the Congress, and was in turn encouraging the Muslims to 'swagger' their new found wealth – via a booming remittance economy with the Gulf – as well as their 'foreign' culture in the form of the conspicuous construction of mosques and new madrasa schools and more orthodox Islamic views (Jaffrelot 1996: 339). The most unsavoury element of the Hindu press coverage was that, in allowing Muslim men to have four wives, they would in effect produce more children than Hindus. This combination of stereotyping (reminiscent of colonial metaphors and tropes concerning the sexuality of Muslim men) and the conviction that the government was pandering to the Muslim economic interests once more emphasised the mobilising strengths of the RSS in organising, through society, counter-demonstrations to defend the majority.

The depiction of a government protecting conservative and traditional Muslims allowed the RSS, and more vigorously the youth wing of the VHP, to support their depiction of Congress pseudo-secularism, a ploy they used to particular effect. They also criticised the Muslims for clinging to conservative religious practices that isolated them from the mainstream of Indian politics and from their own potential as citizens of India (Shourie 1997).

As with accord politics in general, it remains unclear to what extent the prime minister consulted within the Congress and within the cabinet over the decision to introduce the Muslim Women's bill. One minister of state – Arif Mohammed Khan – resigned in protest at the government's extraordinary policy, arguing that the issue was not properly or fully discussed at the cabinet level. Yet Rajiv Gandhi was to follow this move by an almost immediate and blatant appeal to Hindu nationalists, either as an attempt to recover favour or, more perversely, as a 'package deal' to win over religious sentiments all round (Jaffrelot 1996: 378–40). That such a strategy would produce a ratchet effect on ethno-religious demands and expectations, and that there existed a zero-sum game in playing off one community against another, seems to have been entirely underplayed by the political affairs committee and the senior advisors to the prime minister's office (Nugent 1990). On 1 February 1986 – as the Muslim Women's bill was about to be introduced on the floor of the Lok Sabha – the Rajiv Gandhi government was implicated in the unlocking of the gates of the mosque at Ayodhya to Hindu worship, a decision that was viewed as a 'major concession' by the VHP and the radical wing of the Bajrang Dal, which had long been agitating for the cite to be 'returned' to the Hindus, in particular since 1984 (Frankel 2005; Jaffrelot 1996).

As a contested site, Ayodhya goes back to the late nineteenth century (Gopal 1993) and indeed was the subject of legal intervention by the British on the eve of independence, in which it was ruled that the site be denied as a place of active worship for any faith and ordered locked. It was widely believed, and was to be increasingly asserted, that the sixteenth-century mosque had been built on the site of the birthplace of Lord Ram, and that his temple had been demolished under the orders of the Emperor Barbur and the mosque built on the same site as a symbol of Moghul superiority and contempt. One of the many curious features of the mosque concerned the 'incorporation of older Hindu architectural members prominently displayed on the mosque's façade, at a period when the reuse of Hind material was highly unusual, suggest[ing] the patron, Mir Baqi, was attempting to make a general statement of Muslim superiority'(Asher 1992: 30).

Asher goes on to note that these Hindu members are pillars that do not appear to be spolia from a temple dedicated to Lord Ram.[6] Intermittent appeals had been made for the demolition of the mosque , and the restoration of the site to the Hindus for worship since the 1950s, but in 1984 the VHP called for the liberation of the site and brought it to the centre of their ethno-religious mobilisation strategy. This provocative symbol of 'shame' grew in significance as the RSS and the VHP were able to articulate its neglect as symbolic of the Congress betrayal of India's majority community, and attach to it generalised Hindu concerns of being at the mercy of an indifferent, or even hostile, state.

Through a whole series of highly innovative forms of political protest drawing on Hindu ritual and symbolism – above all the Ram yatra launched in 1989 – the RSS and the VHP sought to define a political community of Hindus actively reclaiming their nation, not just as a physical territory, but as an ethno-religious, 'spiritual' space to be privileged over 'outsiders' – namely the Muslims and Christians (Jaffrelot 2004). This process of reclamation involved a careful substitution of the northern town of Ayodhya for the mythological landscape of the Ramayana (Khilnani 1997) and the deployment of Lord Ram as a powerful masculine deity committed to defend – and revenge – the Hindu community (Hansen 1999; Vanaik 1997). In devising ceremonial marches and demonstrations that represented the Hindu nation as a whole (and the resources of the global Hindu diaspora), the VHP was to be incredibly successful in converting Ayodhya into one of the rare Hindu symbols (alongside the cow and the Ganges) that all Hindus could revere and empathise with (Hansen 1999; van der Veer 1994: 138–64). Yet the articulation of religious symbolism was structured by the changes in the state at various levels of the political system, and the ability of political actors to use social and cultural movements to make and articulate 'political' demands (Gopal 1993; Kaur 2005; Mitra 1995).

The Rajiv Gandhi government was complicit in this process, not just in being seen to treat Muslim's differently from Hindus, but in then seeking to manipulate the site of Ayodhya for its own political ends. In 1986, the VHP had petitioned the Faizabad district court to unlock the gates and to allow Hindus to worship. As with previous appeals, the court had followed the advice of the district commissioner and the superintendent of police, both of whom consistently supported the status quo (Frankel 2005; Varshney 1993). However, in 1986 their advice was in favour of the locks being removed, which led to immediate speculation that the government had brought direct pressure to bear on the bureaucracy to influence the courts, a practice that was increasingly common in post-Emergency India and a generalised consequence of calls for a 'committed' judiciary. This was supported by the presence of national and local journalists and TV crews at the site on the day of the ruling. While evidently conceived in the nature of a compromise, the concession merely fuelled the determination of the VHP to press for the demolition of the mosque and the rebuilding of a glorious temple to Lord Ram.

By 1986, the Muslims had responded by forming the Babri Masjid Action Committee, and by 1989 the VHP-RSS had organised a massive national demonstration to converge on Ayodhya with consignments of 'sacred' bricks – Ram Shilas – as a first stage in the construction of the temple. The Allahabad High Court ruled in August 1989 that no parties, groups or persons would be allowed to enter the site, or to change the nature of the status quo, in effect banning the VHP demonstration. Yet such a ban was not to be observed or indeed enforced by the central government, but was to be undermined by further compromises that aimed to allow the VHP to stage its ceremonial demonstration on adjacent land, deemed separate from the area covered by court ruling (Gopal 1993; Jaffrelot 1996). Even when in November the Allahabad High Court sought to extend the area covered under its August ban, New Delhi, in apparent collusion with the Uttar Pradesh government, allowed the VHP to hold the ceremony on a site of their choosing on 9 November (Gupta 1990: 25–50). Furthermore, Rajiv Gandhi chose to initiate his campaign not far from Ayodhya and made ample and exclusive references to Ram's kingdom (Gopal 1993), a symbolic but ambiguous language supposed to convey Congress commitment to the poor and to the building of a prosperous and just India. In using such language, the central government consciously used the language of the VHP. Even references to socialism were now wrapped up in Hindu symbolism. Such a compromise was deemed necessary by the government because of the forthcoming elections and the risk of significant electoral setbacks for the Congress in north India.

Amid the political heat generated by the VHP's Ayodhya campaign, Rajiv Gandhi's authority was rapidly disintegrating at the centre through allegations of corruption, and the Congress faced once again the spectre of non-Congress opposition parties coordinating their attack, spearheaded by V. P. Singh, former Congressman and friend of the prime minister, who had been expelled from the cabinet and then the party for overzealously prosecuting an investigation into the Bofors scandal and for seeking to deal with financial and tax irregularities in general (Nugent 1990).

By 1989 – in the context of a critical election – the tacit and pragmatic support by the Congress for differing religious communities appeared to have been replaced by a wholesale collusion between it and the VHP for the Hindu vote. This emphasis on Hindu majoritarianism was strengthened by V. P. Singh's formation of the Jan Morcha in 1988 (in the company of Arun Nehru and Arif Mohammed Khan) and then, through pre-electoral seat adjustments, the formation of the Janata Dal. V. P. Singh posed a serious challenge to the OBC and lower caste basis of the Congress vote, and threatened to take some Muslim support as well. In the resulting elections, the BJP emerged on 85 seats (having contested 226), taking 11.4 per cent of the vote. Having fallen into line with the VHP–Bajrang Dal strategy, the BJP had to some extent escaped the political cul-de-sac of the late 1960s by deciding to fight the elections alone (on an agenda determined almost exclusively by the Ayodhya issue) while being sufficiently influential enough to compel both the Congress and V. P. Singh to take notice of it. RSS support swung behind the BJP through their opposition to Rajiv Gandhi's support for Muslim personal law and over Bofors.

Although initially excluding the BJP from discussions over seat arrangements, the Janata Dal concluded a series of agreements between the BJP and the CPM not to field candidates against them in specific seats (Kohli 1990: 21–3). Such an adjustment – belying the old fault line of the Janata Party secularists and the former Jan Sanghists in 1978–1979 – made profound inroads into Congress vote, especially in the Hindi belt. V. P. Singh's need to accommodate an increasingly communalised BJP was furthered under-lined when, on forming India's second non-Congress national government, he sought and obtained the outside support of the BJP and the CPM on the floor of the Lok Sabha for what became known as the National Front government. The CPM overcame its objections to too close an association with the BJP on the grounds that it was helping to block the Congress from making a bid for power.

The apparent dramatic gains made by the BJP in comparison with the 1984 campaign was thus symptomatic of the collapse of the Congress, in turn made possible by the ongoing reactions to allegations of corruption, Shah Bano and Ayodhya. Muslims and OBCs voted for the former Lok Dal in Uttar Pradesh and Bihar, while other parties – Telegu Desam from Andhra Pradesh and Janata from Karnataka under Ramakrishna Hedge – revealed the consolidation of regional agendas over national ones. Such fragmentation was again exaggerated through the first past the post system. Congress remained the largest political party, but the Janata Dal/National Front came to power by capitalising on the backward castes and cultivating peasant castes throughout north India, constituencies that had formerly voted for the Congress. The political fragmentation revealed complex regional dynamics between mobilising disadvantaged groups defined by castes: 'polar(ising) political identities between the larger categories of forward and backward castes, but without overcoming caste divisions within the other backward classes and between the other OBCs and the dalits' (Frankel 2005: 628).

Following subsequent state elections in February 1990, the Congress was all but eliminated, managing to survive in just one major state, Maharashtra, with the BJP taking power in Madhya Pradesh and Himachal Pradesh. Through careful choreography, the VHP would be able to main-tain the Ayodhya issue at the heart of national politics, dominating the short life of the V. P. Singh government, which would collapse in June 1991. Even where attempts were made to articulate an emancipatory, 'develop-mental debate' – such as V. P. Singh's momentous decision to implement the Mandal Commission and extend reservations to OBCs – such policies would be interpreted as a response to RSS–VHP strategies to consolidate and reform Hindu society and through this, block the electoral gains of the BJP towards state power.

Like the Janata government of 1977–1979, V. P. Singh had to share power with a series of competitive and notoriously uncooperative individuals, some, such as Chandra Shekhar and Devi Lal, all within his own party. Shekhar's ambition (Shekhar was a veteran Young Turk of Mrs Gandhi's CFSA and one of the few senior members of the party to have been arrested on 25 June 1975) would be one contributory factor in the long drawn-out negotiation over cabinet places, over policy, as well as the fall of the government in June 1991 after Singh's decision to implement Mandal. Unlike the ill-fated Janata experiment, V. P. Singh's decision to form a limited merger premised on the 'outside' support of parties created a more formal and yet open context in which many party disputes could be resolved on an issue-by-issue basis. V. P. Singh proved effective in creating a steering group in parliament to consult over legislation and (with the telling exception of his decision over the Mandal Commission) encouraged open discussions in the cabinet. The onset of coalition revived Parliament's dormant (and arguably ineffective) committee structure and paved the way for significant innovations in cross-party talks that would facilitate the shift from dominant party majorities to coalition government. Under the National Front government, and a necessary caution to the apparent hegemony of the deinstitutionalisation debates, various aspects of the Indian state – the presidency, the Supreme Court, the parliament, the role of the governors – began to separate out and function in different contradictory and at times compensatory ways, as the party system changed their respective weightage within the central state as a whole (Kohli 1990; Varshney 2002).

Yet with the BJP outside the government but still critical to its calculations, and with the Congress still the largest party and on the alert for signs of divisions within the National Front government, there were striking parallels with the dynamics of the Janata period, especially the role the BJP would play in the break-up of the national government. The electoral struggle with the BJP (and within the Janata Dal, between the Lok Dal and V. P. Singh's original Jan Morcha) was set in the context of continuing caste polarisation, and with policies aimed at continuing and deepening state subsidies for agrarian interests deemed to be at the expense of urban interests. Agreements and 'seat arrangements' between these various parties was as opportunist and as cynical as the informal agreements brokered by J. P. Narayan between the former BJS and Charan Singh's former Lok Dal MLAs. Governments were formed premised on the necessity of keeping the Congress on the back foot and not on any specific ideological or policy issues. These sectional interests were themselves overlaid by the continuing agitation over Ayodhya and the fear that it would enable the BJP to create a specific ideological cohesion around the forward castes, and through societal conversion and fragmentation, wrest sufficient backward castes and OBC votes to reintegrate the Hindu vote into a solid electoral wedge through some form form of electoral realignment. In direct confrontation to this, V. P. Singh sought to advance the interests of the lower and backward caste communities – often in conflict themselves over resources – and to accommodate Muslim interests by a significant appointment of mufti Mohammed Sayeed as home minister. In this regard, V. P. Singh was the inheritor of Mrs Gandhi's rhetoric and self-identification with the poor and minorities, if not her actual policy success.

7   *Hindutva* as crisis 1986–2004

The domination of India politics by the Ramjanmabhoomi agitations – from L. K. Advani's announcement that he would lead a procession to Ayodhya in November 1989 to the demolition of the mosque and its immediate aftermath in 1992–1993 – is at on one level, a struggle over how and in what ways the Hindus could be represented as core citizenry of the Indian nation, as both a victim and as an avenger (Kaur 2005). It is about how to depict – and to remind the Hindus continuously – of their shame and their need for a spiritual and national rebirth through strategic acts of violence. Yet it is also about how to use ethno-religious imagery and ideology to make electoral gains and, through the state, to legislate for government and to maintain, again through the state, generic levels of law and order (Brass 1997).

Despite its apparent dramatic success and its apparent affirmation of the importance of religious nationalism in India, Ayodhya and the crisis it provoked disguised a disjuncture within the Sangh Parivar itself, one that had existed from the very moment of the BJS's formation. It would only reappear once the mosque had been demolished: A disjuncture between a political party seeking election and a series of social movements committed to the long-term renaissance of Hindu society.

The irony in trying to conceptualise the tension between the political, social and religious elements of the Sangh Parivar formation is that the rapid and dramatically uneven caste and class mobilisations that were taking place throughout the subcontinent initially obscured it, as did a series of political gaffs by the central and state authorities. These provided the highly unusual circumstances in which these two quite differing strategies could work together. What Jaffrelot calls, rightly, 'the impossible assimilation' was the product of ongoing fragmentation and regionalism that allowed the ethno-religious mobilisation strategies of the RSS–VHP to momentarily dovetail with the political, programmatic logic of the BJP. It was also the product of a state elite which was too weak to challenge a movement that seemed to them to have very real electoral significance.

Yet in 1996, three out of four people in India were voting for secular parties, and more than 21 per cent of the electorate voted for state parties, and only 70 per cent of the vote went to officially designated national parties, of which only eight qualified (Frankel 2005). These figures were roughly the same for the 1998 elections that produced the BJP-led National Democratic Alliance (NDA) that would become India's first national coalition to go a full term in office. Throughout, the BJP's subsequent success – as much as the Congress in 2004 – lay in the strength of its coalition partners and the BJP's reliance on regional or regionalised political formations, but this forced it to dilute or indeed set aside the wider ethno-religious agenda of the RSS–VHP, with its use of violence and the threat of violence, which could necessarily sustain an electoral momentum (Hansen 1998). The BJP's highest share of the national vote was in 1998: 25.6 per cent, down to 22.2 per cent in 2004. It is tempting to suggest, from the vantage point of 2006, that ethnographic and sociological writings on Hindu nationalism – as well as the debates on over whether Indian nationalism was 'indigenous', 'authentic' or indeed merely derivative – have exaggerated the role and essentialism of religion, or rather, seriously underplayed the extent to which, once co-opted by the state and a rather surprisingly old-fashioned emphasis on electoral systems and party politics, these symbolic processes are exposed as being weak or contradictory on policy and unable to mediate the complexities of state–society interactions once they become institutionalised (Brass 2003; Brass and Vanaik 2002). Notwithstanding the ongoing societal mobilisations of the RSS (to be discussed below) and arguments that the Sangh Parivar have insidiously transformed the positions of its political coalition partners and their social imaginings, scholarship in the mid 1990s underplayed the programmatic weaknesses once in power of religious nationalism and the highly charged, transitional circumstances that allowed its use of instrumental violence and intimidation to hide this very weakness.[1]

Part of the disguise was the existence of the mosque itself and the BJP's electoral position in 1990. While it existed, the mosque remained as a potent symbol of Muslim oppression, while also forcing the Muslims to identify themselves with what Arun Shourie referred to derisorily as a 'defunct place of worship' (Shourie 1997). In conditions of huge political uncertainties, the dispute forced opposition parties in and out of government to adopt a cynical expediency towards the RSS-VHP, both in allowing demonstrations and processions to take place, and in prevaricating over how best to deal with the various and escalating VHP demands, without seeming to alienate Hindu sentiment. Hindu sentiment – as a religious constant – was more in the minds and calculations of the political elites than it was present in society, where it was often disputed, transient and localised. Such passivity by the central state in the face of the Ram Shila procession in 1989 would have been unthinkable ten years earlier. This type of mobilisation – and the communal violence it generated in its wake – resulted in tangible electoral gains for the BJP in 1989, despite its reticence in initially embracing the sadhus and its slowness in moving into tandem with the RSS networks.

Identified rightly as rituals of confrontation as well as rituals of mobilisation in which – through processions – fragmented social structures are re-imagined as an undifferentiated community (Jaffrelot 1996: 392), such useful sociological perspectives undervalued the significant fact that *the state gave way* and amplified the impact of *Hindutva* immeasurably (Freitag 1989; Kaur 2005). Successive governments from Mrs Gandhi through to her son, to both the National Front and Congress-I governments that followed, failed to deal adequately with the VHP, and instead of preventing the orchestration and incitement to political violence, were drawn into it, a process of convergence that brought huge gains to the BJP.

By August 1990, V. P. Singh had shifted from a tactical reliance on the BJP to utilising a key legacy of the Janata Government – the Mandal Commission – to undermine the RSS communalising strategy by implementing the commission's recommendations for setting aside 27 per cent of government posts and posts in public sector enterprises for members of OBC communities. Calculated as part of a wider commitment to a populist socialist strategy, the decision aimed to construct a powerful block of support amid the poor, far more dramatically than the language and rhetoric of Mrs Gandhi's *garibi hatao*. It dared the BJP to condemn the move and risk alienating itself from the material aspirations of just over 50 per cent of the Indian population, but it created serious divisions within the Janata Dal party, especially between Devi Lal and V. P. Singh. It also sparked a series of caste wars between non-elite dominant landowning castes and the levels of sudra castes below them, and widespread urban unrest and protests amongst forward caste, middle class families (Frankel 2005). At the time of the announcement, 'the VHP had just gathered together 5,000 sadhus at Vrindavan to announce that thousands of kar sevaks would depart for Ayodhya from September onwards' to push forward with the demand for the building of the temple (Jaffrelot 1996: 414). Led enthusiastically by L. K. Advani, he later stated in an interview that although long contemplated, it was nonetheless an ideal counterstrike to the National Front government's attempt to fragment the Hindu votary along caste and class lines (Jaffrelot 1996: 414–15).

The role played by the VHP-BJP 1990 processions and their role in the fall of the V. P. Singh government are well documented (Gopal 1993; Varshney 1993). The Rath Yatra – chariot procession – maintained pressure on the National Front government and sustained the momentum of what was seen by elements of the RSS as a significant social breakthrough for Hindu reform generally: namely the creation at Ayodhya of a 'national' centre for the Hindus, both in terms of the re-imaged Indian landscape as Ram's kingdom, but also in terms of constituting an ecclesiastical centre to the Hindu faith (van der Veer 1994). The procession, starting out from Somnath, was popularly received by demonstrations celebrating elements of the Ram myth in dances and song, or in presenting blood for their anointment. As evidence of the total collapse of the national secular norms, Jaffrelot notes that 'for the first time, a political leader used propaganda of an overtly Hindu nationalist character throughout eight of India's states' while also conceding that such an 'innovation was made possible thanks to the political situation' (Jaffrelot 1996: 417). This situation again counselled both the Congress-I and the Janata Dal towards caution and appeasement.

V. P. Singh tried in vain to prevent his reliance on the outside support of the BJP from undermining his attempts to maintain the status quo of Ayodhya and preserve the mosque, while the Congress ensured that no obstacles were placed in Advani's way when it passed through the Congress-run states of Maharashtra and Karnataka (Gopal 1993). Only in extremis, when it was clear that Advani would not be stopped and that the National Front government would collapse did V. P. Singh authorise the chief minister of Bihar, Laloo Prasad Yadav, to arrest the leader of the BJP before he crossed into Uttar Pradesh on 23 October 1990. In response, the BJP withdrew its support in parliament and launched a national agitation.

Communal violence had already taken place before Advani's arrest, especially in the wake of a series of flame processions aimed to symbolise the awaking of the devotees of Ram,[2] and in specific flashpoints where the Rath Yatra passed through Muslim areas. As the Uttar Pradesh government carried out mass arrests – 150,000 under so-called protective custody orders – communal rioting was reported in 26 separate locations (Home Ministry Report 1992–1993, 1993–1994). In Ayodhya itself, kar sevaks forced down the cordon around the precinct of the mosque and were able to plant a saffron flag on one of the three domes before the Border Security Force were deployed to clear the area. The consequent shooting and tear gassing of ker sevaks generated a whole new set of images and myths that would be used as propaganda – graphically presented in a short video, and finding their way into audio cassettes and video 'chariots' – to maintain the Ayodhya Movement and to shame and criticise the government. The official records show that six people were killed, but the VHP was to claim as many as a hundred martyrs, the remains of which were to be paraded and processed throughout India in a rather extraordinary series of 'blood and soil' demonstrations.

V. P. Singh resigned, to be replaced for a brief stint by Chandra Shekhar, who was eventually brought down by the withdrawal of support by the Congress-I. Fresh national elections in 1991 produced, amid the ongoing fragmentation of the Janata Dal and the assassination of Rajiv Gandhi on the eve of the final leg of polling, another hung parliament. During the campaign, the tensions between the BJP and the VHP-RSS (and increasingly the undisciplined volunteers of the Bajrang Dal) began to emerge. In the electoral bargaining that followed the results, a Congress-led minority government emerged under Narasimha Rao,[3] but the BJP had further advanced its parliamentary support and was now the second largest party in the Lok Sabha. It had also taken control of Uttar Pradesh itself as the caste–class basis of the Janata Dal fragmented into mutually hostile constituencies (Frankel 2005; Leiten 1994). In such close proximity to state power at the centre, the BJP was compelled to display itself as a responsible political party, both with reference to law and order and the prevention of communal violence, and to wider policy remits on matters of foreign policy, constitutional government and, by 1993–1994, ongoing economic reforms (Hansen 1998).

In many ways, the extent of the BJP's electoral success in urban metropolitan India in 1991 underlined the contradictions between it and the rural or small town austerity of the RSS, with its emphasis on a disciplined, non-consumerist lifestyle. Alienated by both the Congress and the Janata Dal, the middle classes took to *Hindutva* both vicariously and selectively – attracted to its emphasis on order and discipline and on its forward caste pedigree (*India Today*, 15 February 1993). Middle class professionals, active and retired civil servants, judges and members of the military for the first time became open members of the BJP, something that would have been unthinkable earlier, even during the 1977–1979 period. Yet the extent of the BJP victory – that many saw as constituting a form of breakthrough or electoral realignment – logically implied to party workers and volunteers that the momentum for the Ram temple would continue, and that the party would direct its political fire against pseudo-secularism and the position of minorities in particular. Yet the violence associated with the VHP and with sadhus and members of the Bajrang Dal was a growing concern to the party, especially as the agitation for Ayodhya was increasingly referred to the authority of the courts. The new BJP state government in Uttar Pradesh sought to bridge the gap between the diverging demands of party and movement by acquiring land adjoining the Ayodhya site, and despite writs issued to challenge it, land clearance took place as in preparation for the construction of a temple close to, but not on, the site of the mosque. The Allahabad high court upheld the purchase of 2.77 acres by the Uttar Pradesh state government, but ruled that no permanent structure could be built on it (Jaffrelot 1996). The Supreme Court later reiterated this ruling.

From November 1991 until December 1992, amid an attempt by the Rao government to resume negotiations between the VHP and the Babri Masjid Action Committee, and ongoing interventions and referrals to the courts, the differences between the BJP government and the VHP proliferated. Ironically, neither Muslim nor Hindu organisations saw the Rao government as 'neutral'. The VHP announcement that construction of the temple would start on 6 December 1991 appeared to have caught the BJP state government off guard, since Kalyan Singh had reassured the Supreme Court as early as November that no further demonstrations would be allowed. This statement sparked off a further round of meetings, and a compromise – this time between member organisations of the Sangh Parivar and a BJP state government – opened up the possibility for a 'symbolic' kar seva to take place on nearby undisputed land – a strategy reminiscent of Rajiv's deal with the VHP in 1989. Yet as a precautionary move, New Delhi deployed paramilitary forces at both Ayodhya and Faizabad, despite protests from the chief minister that this was 'anti-federal'.

The involvement of the central BJP leaders – L. K. Advani and M. M. Joshi, and indeed their presence at Ayodhya – was to complicate issues over responsibility and intentionality in the wake of the mosque's destruction on 6 December. Jaffrelot, citing a variety of sources and having interviewed some of the key actors involved, effectively argues that while it is impossible to blame the demolition squarely on the BJP, sufficient evidence exists to reveal that the demolition was planned and even rehearsed (Gopal 1993; Jaffrelot 1996: 454–56). Both Advani and Joshi were present when kar sevaks, apparently undisciplined by RSS cadres present, swarmed over the mosque and, by early evening, reduced it to rubble. Circumstantial evidence suggests that the VHP–RSS planned the assault, with the BJP leaders being taken on board later on. Whether the leaders of India's official opposition party protested or not, the events strengthened the image that sadhus and kar sevaks were stronger within the Hindu nationalist movement than the disciplined forces of the BJP – that, in simple terms, their volunteers were 'out of control'.

For a movement whose ideological orientation stressed discipline, and the emergence through self-control of a new Hindu, this was a damaging revelation, no matter how hard they tried to project the blame onto a heartless and incompetent national government. For a political party whose voters saw it as both protecting caste and class privileges, and in some cases, the sanctity of property, the consequences were equally disastrous. The aftermath to the demolition of the mosque revealed that the electoral surge in favour of the BJP from 1989 to 1991 had been brought about through the fear of the rise of the OBCs. As Barbara Harriss-White notes, in referring to the obvious affinity between members of specific social class, the rise of the BJP and *Hindutva*, 'while some members of these classes may support the casteist and economic–nationalist project of swadesh, and the ambivalently casteist and Hindu supremacist project of Hindutva, neither swadeshi or hindutva are necessary – they are simply consistent with their interests' (Harriss-White 2003: 68)

Blame did fall on the handling of the affair by the Rao Congress government in Delhi, in its slowness in deploying central forces or in anticipating the difficulties that the BJP government would have in policing its own cultural–religious wing. In response to the outrage, and to widespread communal violence throughout north India, especially in Bombay (and disturbances in Bangladesh and Pakistan), the BJP leaders, including Kalyan Singh, resigned (Jaffrelot 1996; Varshney 1993). The central government dismissed from power all of the BJP state governments on the grounds of shared moral responsibility, a move reminiscent of the post-Emergency moves by the Janata government in 1977, and one carried through on a very similar premise: of shared guilt and collective responsibility. The Rao government committed itself to reconstructing the mosque. The dismissed governments petitioned the Supreme Court to rule that such a use of Article 356 was 'gratuitous', and in a landmark case, the Supreme Court upheld such a view and, with the exception of Uttar Pradesh, ordered the reinstatement of the former BJP governments. The Congress government momentarily arrested Advani and Joshi on the grounds of inciting racial hatred, but they were later released.

The RSS and the Bajrang Dal were also banned, but 'even after the demolition of the Babri Masjid, the central government's policy towards the RSS–BJP–VHP combination oscillated between firmness and conciliation' (Jaffrelot 1996: 465). The central government's commitment to rebuild the mosque soon concerned Congress advisors as needlessly provocative. In comparison with earlier bans, there is little evidence that this one was carried out with either thoroughness or conviction. It was also complicated by the intervention of the courts, in which high courts ruled that the Bajrang Dal and the RSS were cultural organisations, and as such could not be banned for political purposes. In Uttar Pradesh itself, despite remaining under President's Rule, the appearance of a makeshift temple on the site of the former mosque implied that the centre was complicit in the continual erosion of the earlier robust response to the demolition. Arguments advanced by 'secularists' for direct action against the RSS even struck many in the BJP as being un-democratic: how could a movement be outlawed that, in terms of volunteers, claimed to be the largest organisation in India? Such a view insidiously aided the growing expectation that, whatever their complicity in the events of 6 December, the BJP were surely poised to form a government in Delhi soon?

Furthering its strategy of appeasement, the government made use of another Allahabad ruling striking down the earlier land purchase of the Kalyan Singh government and sought to refer the whole matter to the Supreme Court (Jaffrelot 1996; Varshney 1990). It also tried a more sophisticated route through engineering a split within the sadhus to weaken the VHP – and indirectly, the Sangh Parivar more generally – by getting Hindus to denounce the organisation, and utilising their own sadhus and religious leaders to denounce the VHP as 'un-Hindu'. In seeking to legislate solutions at the centre, the Rao government was exposed to the vagaries of coalition government and the failure to construct a broad front against the BJP. Attempts to separate religion from politics mirrored the same difficulties experienced during the 44th Amendment and the then proposed attempts to define 'secularism'. No agreement emerged, and a separate attempt to amend the Representation of the People's Act also failed, with the government proving incapable of getting the Janata Dal on board. In the debate in the Lok Sabha, the Janata Dal accused the Congress of seeking to legislate as a substitute for political will.

Nonetheless, the demolition of the Babri Masjid, whether or not it was a accident or even a unanimous decision, completely transformed the 'terms of the strategy of ethno-religious mobilisation that the RSS-VHP-BJP combination had been pursuing for almost a decade' (Jaffrelot 1996: 457). The degree of communal violence, and the dismissal of BJP governments by the centre, also redirected the concerns of the Sangh Parivar, as did the embarking by the Rao government on a strategy of neo-liberal reform which stressed socio-economic policies at the expense of ethno-religious nationalism. These policies would, through the 1992–1996 period, expose further differences among the Hindu nationalists within a complex array of state and society interactions.

In 1993 five state election campaigns were held in the wake of the Babri Masjid demolition. The seemingly determined position adopted by the Congress-I government in New Delhi placed the BJP at a disadvantage. Its national leadership sought to widen its socio-economic and national appeal, distancing itself from the VHP, and attack the Congress government on the basis of its foreign policy, its economic liberalist strategies and ongoing corruption. The removal of Ayodhya fragmented a clear focus on Hindu grievances, and also the Hindu nationalists faced the new variable of having their time in office – and the extent of policy success – judged by a now highly mobilised and critical Indian electorate with a tendency to punish incumbency. With power came the associated difficulties of factionalism, disputes between party organisational wings and MLAs, and allegations of malpractice (Jaffrelot 1996: 494–521; Manor 2005).

The articulation of swadeshi to criticise the government's approach to the IMF and liberalisation of the economy was problematic and mapped unevenly onto the BJP's emerging urban constituency (Harriss-White 2003: appendix 1). It implied a return to economic nationalism, based on a strict form of state control and small-scale goods. It promised the way back to some form of permit license Raj, and as such proved unappealing to a middle class-high caste vote bank, which saw liberalisation as a way of acquiring new consumerist products, and of expanding their profits through global markets.[4] Economic policies revealed the huge gap between the bastion of BJP support (the 'pro-rich') and the party's need, at the same time, to break out and obtain the support of backward and other backed castes through more inclusive statements on development and reservation policies.[5] The growing political consciousness of the OBCs was probably the main obstacle that faced the BJP, both in terms of its socio-economic policy and over its association with the RSS and the ongoing social, sangathanist strategy to create a reformed Hinduism that privileged forward caste, Hindi-speaking, north Indian Hindus.[6]

In opposing the Rao government at the centre, the BJP had more scope to articulate the symbolism of *Hindutva* by tabling a series of debates over the need for a uniform civil code, in effect calling for the government to legislate to remove customary personal law and treat all citizens as equal. Two separate attempts were made in 1993 alone to pass private members bills. This proved an effective way of challenging the government over the legacy of Shah Bano in the wake of the unsuccessful attempts by the Rao government to effectively legislate against ethno-religious nationalism. Grounding their debates in the secularism of Article 44, the BJP successfully made a great deal of the obvious bind facing the Congress. As a party, the Congress was in favour of legislating on the basis of Article 44, but could not afford to agree openly with the BJP in the wake of the demolition of Ayodhya.

For the BJP, calling for the implementation of a uniform civil code was a way of confirming its secular credentials and its support for the founding fathers of the constitution. On many issues, such as Kashmir, Pakistan, globalisation and development, the BJP seemed unclear on some its policies or in active disagreement with the RSS over the 'impossible assimilation' between the long-term, societal plans of the RSS and the short-term, pragmatic responses required to win power and to maintain it in a democratic fashion. In the states, the BJP-led administrations that did emerge were cautious about their relationship with the VHP, not just as a momentary illegal organisation, but one that was seen to be too volatile and single-issue based to be of any further electoral use (Hansen 1998).

By 1996, amid the extraordinary allegations of the *hawala* case (Frankel 2005), the Rao government went to the country for India's eleventh general election under an unprecedented cloud of corruption, which linked Delhi's handling of economic liberalisation to the mass looting of India's public funds (the CBI exposed a network of bribes from the prime minister down to senior bureaucrats working in some of the states). In comparison with earlier corruption scandals, especially those of the Indira Gandhi era, the scale of the *hawala* case was unprecedented. The BJP mounted a cautious campaign in which it reorientated itself still further from an emphasis on the ethno-religious symbolism of *Hindutva* embedded around the now defunct Ayodhya. At the party conclave in Sariska in 1994, it signalled a rethink on changes in economic policy (diluting swadeshi in favour of liberalisation) and in its approach towards minorities (Mitra 2005: 80). The only remaining commitment to an overt *Hindutva* agenda was the commitment to the passing of a Uniform Civil Code; this was addressed repeatedly by the leadership in the election campaign.

The 1996 election campaign dramatically confirmed the regionalising trends within Indian politics, with the Congress-I's share of the vote collapsing and with the party taking just 139 seats out of a house of 543. The Janata Dal continued to fragment into regional and state factions, 'a victim of its own mobilisation strategy' (Frankel 2005: 691) while the BJP was returned with 161 seats, and for the first time, emerged as the single largest party. The peculiarities of the Indian electoral system produced this result for the BJP, despite the fact it was well below the Congress in the share of the popular vote (20 per cent, compared to the Congress's 29 per cent), and despite the fact that the Congress remained the only political organisation that could mount an election campaign across India as a whole, contesting 529 seats as opposed to the BJP contesting 471. The Indian electorate fragmented dramatically. 'More than 21 per cent of the votes were cast for state parties in 451 constituencies and resulted in victories for them in 127 seats. Over 3 per cent of votes polled when to "registered, un-recognised parties"...6 per cent of the electorate voted for one of 10,635 independents' (Frankel 2005: 691; McMillan 2005).

In the attempt to form a government, the BJP sought to create electoral alliances with a number of regional parties, explicitly seeking to placate concerns that the party had a hidden political agenda with regards to the minorities and the Muslims in particular. Congress-I and a series of self-confessed 'secular' parties sought to isolate the BJP which, as the largest party, was called by the president to form a government and then seek a vote of confidence in the Lok Sabha. The decision by the president was contested by those who sought, ineffectually, to portray the BJP as an 'illegitimate' party that would, through *Hindutva*, challenge the basic structure of the constitution by undermining the state's commitment to secularism. There were also unsubstantiated claims that the party would seek to install a presidential system or would, in collusion with the RSS and unelected and unaccountable elements of the Sangh Parivar, create an unprecedented law and order situation. Throughout the English language press, as in 1975, there was much talk and debate over fascism.

Responding to the national outcry over the destruction of the Babri Masjid, Vajpayee had reiterated the BJP's commitment to the Indian constitution and, amongst the senior leaders of the Sangh Parivar, he alone had come the closest to apologising for the event which he described as the 'the biggest mistake' ever made by the BJP. He was depicted and projected as the 'moderate'[7] who would, in government, steer the party towards 'common sense' policies. The BJP-led coalition of 1996 lasted an infamous 13 days, and resigned without facing what was, in the face of a tactical agreement between the Congress-I and a 15-party coalition, a foregone defeat before the lower house. Congress-I consciously set out to underpin secular forces and confront the communalists, but in a language that obviously underplayed its earlier collusion and which quickly petered out into an attempt to mediate the dispute as opposed to end it.

Despite Vajpayee's attempts to project himself as an elder statesman, drawing on his reputation as a parliamentarian and as a former foreign minister, concerns expressed in the national and foreign media concentrated on the RSS, which by 1998 claimed to be the largest voluntary organisation in the world, with over 30,000 branches and, through the active involvement in the wider Sangh Parivar, 140 front organisations penetrating into almost every aspect of Indian daily life (Frankel 2005: 727). The RSS and its relationship with the VHP and the Bajrang Dal was largely perceived as an inappropriate one for a democratically elected government, given the degree of consultation that took place between them. Frankel noted that 'there was no precedent in Independent India, or any other democracy, for a ruling party which acknowledged the ideological authority of a "parent" body claiming to be a cultural organisation with no political accountability (Frankel 2005: 728).

Significantly though, in comparison with the 1979–1980 controversies over 'dual membership', the debates over the links between the RSS and the BJP were not so prominent, and this time largely failed to undermine the BJP's claims to be a serious political party. The contrast is instructive. In part it reflects the normalisation of the RSS as part of the Indian socio-political landscape and the ways in which its language about *Hindutva* and Muslims had become part of a communalised common sense (Jaffrelot 1996: Robinson 1997). This had been helped by a series of court cases that had ruled over the meaning and scope of *Hindutva*, judging them to be in keeping with 'a way of life' and not an explicit reference to religion, while other rulings had confirmed the 'cultural' nature of many elements of the Sangh Parivar and which had 'unbanned' the RSS and the Bajrang Dal (Robinson 1997: 1579). It is certainly the case, however, that the fiasco of the 1996 bid taught the BJP important lessons over the need for pre-electoral pooling (McMillan 2005: 13–35). Between 1996 and 1998, the party resumed (with the tacit consent of the RSS) the shift towards a policy of 'integral humanism' as the guiding light of the BJP's broad commitment to government, a slogan that echoed the BJP's position throughout the early and mid 1980s. In order to win power, what was required was a conscious dilution of the *Hindutva* agenda and an attempt to 'engineer' caste–class coalitions with OBCs and non-Hindi-speaking shudra castes in the south, while retaining their core voting areas, namely the north and west. The dilemma of the party in 1996 had been neatly summarised by an editorial in *The Hindu*:

> Much as the leading BJP protagonists, at their persuasive best, tried to portray the BJP as committed to the ideal of a pluralist, multi-religious, multi-lingual and multi-ethnic country, as long as the BJP remains anchored doctrinally to the dominance of Hindutva...the concept of a homogenised India in which the minorities will have to speak the language of the majority – it will continue to be perceived as a wolf at the door.
>
> (*The Hindu*, 1996)

At the end of 1996, this dilemma seems to have been well understood by the BJP as a party that increased its efforts through the 1996–1998 National Front government to find pre-electoral allies (Hansen 2001). In 1998, the BJP campaigned on the basis of a series of pre-electoral agreements with 14 parties (some of them extremely small), one of the most significant allies being AIADMK in Tamil Nadu. In consultations through a pre-electoral steering group, the party had agreed to drop its most contentious electoral pledges, viz. the adoption of a uniform civil code, the abolition of Article 370 for Kashmir, the building of the Ram temple at Ayodhya and the abolition of the minorities commission. As part of a comprehensive strategy of pre-electoral agreements, the BJP contested less seats than in 1996 (388 compared to 471) and yet increased its number of MPs to 182, as well as its overall share of the popular vote, up to 25.6 per cent, and closing the gap on a Congress-I that dropped to almost the same share (Hansen 1998). When, as in 1996, the Indian president called for Atul Vaypayee to form a government, he did so in the knowledge that it would survive its first vote of confidence.

The onset of the first BJP-led national coalition government – the National Democratic Alliance – met a cautious response in the Indian press, but this response was less strident and more accommodating that the coverage of the 1996 bid for power. Surveys and opinion polls across the intervening years had revealed the extent to which, amid the continuing collapse of national parties, it was seen as inevitable that the BJP would 'have to be given a go' at office (*Economic and Political Weekly*, 7 March 1998: 491) Significantly, the debate was not about the commitment of the government to the poor or the extent to which it would re-engage with anti-poverty programs and support subsidies (however targeted), but whether it would threaten secularism and the minorities. The *Economic and Political Weekly* noted that even 'a cursory glance at the post-election arithmetic shows that taking into account every possible permutation and combination, any government that is likely to be formed will be preoccupied, above all and mostly, with its own survival. It will naturally lack any ideological and programmatic coherence' (*Economic and Political Weekly*, 1998).

And indeed the first thing that the party had done was to put Vaypayee on national television where, with a picture of Nehru in the background, he had pledged loyalty to the principles of secularism and sought to assure the minorities that there would be no changes in their status and to their rights. The speech was choreographed to downplay the imagery of an ideological coup, and reiterate the continuities of government. The impression of business as usual was reinforced later when, amid concerns over RSS reactions, the new finance minister confirmed the government's continued support for the liberalisation strategies adopted by the Rao government and sustained by the National Front government between 1996–1998 (Adeney 2005: 97–115; Mitra 2005). In fact the government was to last a year, after which it was to collapse over difficulties with its AIADMK partner. Its short period in office had been marked by paralysis within government – especially at the cabinet level – and difficulties with managing divergent – and diverging – aspects of state power. The decision to test a series of nuclear devices in 1998 was the one policy that lent support to *Hindutva's* projection of India as a great power and to claim, indeed demand, international attention (Chiriyankandath and Wyatt 2005: 193–211; Hewitt 2000).

In the wake of the Kargil incident in 1999, where Pakistani irregulars laid siege to the shia town of Kargil (approximately 12 miles inside Indian-administered Kashmir), national elections were postponed and Vaypayee headed up a caretaker government. His prosecution of the war, and the ability of the BJP to support a nationalist cause, partly explains the return of the BJP-led national coalition in 1999. Other additional factors concern the consolidation of the 1998 alliance-seeking strategy, and on this occasion, the publication of a pre-electoral program for government, the so-called Agenda for National Government (Adeney and Saez 2005), which further marginalised *Hindutva*. The National Democratic Alliance this time consisted of 22 parties within the government, and two parties supporting it from outside. Many of these parties were the product of the further regionalisation of the Indian electoral landscape and drew their support from constituencies either indifferent or hostile to a wider *Hindutva* agenda. Congress-I, in failing to recognise that the one-party dominance and the electoral-wave phenomenon were things of the past, flouted the necessary logic of pre-electoral agreements and appeared to many to have finally bankrupted itself of any organisational, ideological or even pragmatic claims to government, let alone its historic claims to political hegemony (Frankel 2005: 715).

The political landscape of state–society relations between 1999 and 2004 appeared, in contrast to the Congress years, to have changed beyond recognition. It raised questions not only as to 'whether a new pattern of Indian governance was emerging' (Adeney and Saez 2005: 3) but to what extent there had been an ideological shift toward Hindu nationalism. A dramatic rise of communal violence, implicating organisations within the Sangh Parivar, and the association of the BJP with the destruction of the Babri Masjid in 1992 created fears that the rise of Hindu nationalists would create, if not a specific and peculiar version of Indian fascism, then a majoritarian form of democracy that would be intolerant of diversity, indifferent to the poor and increasingly authoritarian (Hansen 1999: 7–11). Yet the realities of the BJP in power – and the BJP-led National Democratic Alliance becoming the first coalition government to effectively exorcise the ghost of the Janata government and go the full five years in office – are complex, and are not easily read into a neat, ideological interpretation of events, or through a discourse dominated by religion and the Muslim other. (see Mitra and Singh 1999; Mukhopadhya 2004). In more narrow terms, the rise of the BJP as a political party can only be explained in the context of the collapse of the Congress and the strategies of electoral mobilisation that the Congress had unleashed in order to stay in power.

The BJP came to power in 1998 not as a single party, wedded closely to its parent body the RSS and the various societal fronts of the Sangh Parivar, but in coalition and in active (and from 2002, growing) disagreement with the RSS. Having successfully argued that it was compelled to accommodate India's federal polity in order to win power, such pragmatism arguably became a way of sustaining it, to the evident disdain of the RSS.[8] Although there is evidence that the RSS combine has insinuated itself within Indian society and changed the nature of political debate (the so-called discourse of common sense views on matters as diverse as the poor, the minorities, the role India should play in the world and what sort of state would facilitate this role) (Anand 2007), it is evidently clear that the BJP has often not been in a position to legislate and implement policy conducive to a *Hindutva* project. Moreover, while there is evidence that the BJP has, through the patronage of state power, appointed a whole series of actors including high court judges, university vice-chancellors, police chiefs, education committees and governors, the state–society arena has on the whole proved too complex and dynamic to homogenise the social order and through this, transform the logic and experience of state power. Politics since 1992, and in particular since 1998, have revealed that minorites have often been more empowered and active than at first assumed (Mitra and Singh 1999). Where the BJP-led coalition has sought to implement unadulterated *Hindutva*, its path has often been checked.

A useful example of this comes from 1998 in what was, in effect, a much more sophisticated rerun of the Janata textbook controversy. In October, M. M. Joshi, then human resource development minister in the national government, convened a conference to look at the education curriculum in schools. Attending the conference were a series of educational ministers from the state governments, many of them formal allies of the NDA government in New Delhi (Lall 2005). Among the papers under discussion was a special report by a 'specialist' organisation – the Vidhya Bharati – which was an educational affiliate of the Sangh Parivar and the RSS. The paper recommended a number of changes to the curricula in order to reflect 'Indianised, nationalised and spiritualised' issues; in effect, the ideological message of the RSS. The report also proposed amending Article 30 in order to undermine the rights of minorities to educational autonomy (Ruparelia 2005: 35). Such a move led to widespread protest amongst NDA allies and from the opposition and the press, and a swift U-turn by M. M. Joshi.

As Ruparelia points out, however, in other cases the success of the BJP in getting over an ideologically coherent project has been more successful, but the nature of this success bears close examination. Since the emergence into power in 1998, incidents of violence against the minorities (especially Christians) had constantly provoked debates (some of them remarkably indifferent), but despite methodological issues of defining and recording communal violence, the incidents of communal violence declined throughout the NDA's period in office (Mitra 2005: 95; Wilkinson 2000: 767–91). Only in one specific event – in Gujarat in 2002 – has there been an apparent convergence between societal organisations, politically instigated communal violence and the consolidation of a Hindu votary supporting the BJP en masse (Mitra 2005). What happened in Gujarat was exactly the sort of assimilation of state and societal power that the critics of the BJP expected the party to do once it had entered the portals of national power – the very thing many argue that its coalition partners and wider societal vigilance have prevented it seeking to orchestrate once in command of the central state. While possible as a crystallisation of state–society power at the local and state level, and in particular Gujarat, long perceived as a laboratory of *Hindutva*, at the federal level this sort of strategy remains almost impossible (Yagnnik 2005). This one incident, and the failure of the wider federal government (with the notable exception of the Supreme Court), is instructive, both for revealing the specific dynamics of Gujarati politics as well as the wider context of Indian federalism and democracy (Brass 2003; Hinnells and King 2006). Both the strength and the weakness of Indian democracy lies in its mobilisation of the poor to vote, and the reactions such direct participation has brought to political elites. The BJP benefited from this and from ethnic outbidding by the Congress, but it has yet to consolidate itself within the central state, or through the RSS, generated the necessary all-India wide sense of 'crisis' to bring about the convergence between its societal *Hindutva* agenda and the BJP's strategy of asserting political dominance. Seeing the Sangh Parivar as a coordinated collective strategy for power missed the nuances of these contradictions (Jaffrelot 2005; Noorani 2000).

Coalition politics has contained the BJP, even as it has brought about some convergence between coalition partners and the wider aspects of the Sangh (Adeney 2005: 98). The change in state–society relations witnessed in India since the 1970s involves not just the co-optation by the state of emergent societal interests and changes within and between what have been called 'the dominant coalition', but also changes within the state itself and the social imaginings of the state within specific elites. Together, these have been significantly under-theorised, in part because even relatively sophisticated approaches to the Indian state have comprehended it in monolithic, isolated terms of reference. The cause and consequence of the electoral decline of the Congress, and the subsequent if less hegemonic rise of the BJP, requires an understanding of the state

> [t]hat encompasses complex layers of political institutions that operate in a variety of ways in different contexts.... Indeed, instead of talking about the Indian state as a monolithic entity, it is preferable to talk about the Indian state as the sum total of myriad forms of political institutions in order to develop a more comprehensive understanding of the sub-continental kaleidoscope of their political institutions and their immediate social environments.
>
> (Bates and Basu 2005: xiii)

Since its return to government, the Congress has, under the dual leadership of Sonia Gandhi (party president) and Mohan Singh as prime minister, operated in a complex and dynamic state–society relationship that implies coalitions are now the 'governmental' norm – a form of administration that, at key moments, the Indian state has moved towards since 1967. Is this system one that will close the gap between political mobilisation and the experience of democratic governance that has, for so long, created and defined the pathologies of the world's largest democracy? (Varshney 2002).

Conclusions

It has been one of the basic tenets of this book that, given the historical relationship between the Indian state and the Indian National Congress Party, the end of Congress dominance would invoke a series of crises, contingent on the Congress collapse as institutional decline and transformation removed the party as an informal mechanism for directing and coordinating state–society power. This was not just with reference to the judiciary, the executive and the legislature, but also federalism generally: how politics had been defined, and in what way decisions were taken and then implemented or not. As India entered a decade of coalition, the various institutional aspects and levels of state–society opened up to complicate the processes of governance and the operationalisation of democracy.

The Congress collapse was determined by political mobilisation and the party's response to this, namely the switch to an overtly authoritarian populism, and then in the wake of the collapse of the Emergency, the improvised, tactical alliance with selective communal appeals and ultimately, a rendezvous with *Hindutva*. Through the peculiarities of the Congress system and its complex links with a caste–class society, the uneven mobilisation triggered by the collapse of the Emergency dramatically increased the participation of the poor, but fragmented political accountability and allowed a party – and parts of a wider ideological project – that was, elitist, intolerant, and anti-poor to come to the fore. Thus, in rather stark normative terms, the rise of the BJP and its overt association with *Hindutva* are defined here as a pathology of a moribund political system and a massive failure of political leadership, primarily and historically within the Congress.

Just as the Emergency was made possible by the constitutional arrangements of the Indian state and the peculiarities of the Congress system that breached the weak balance of powers at the centre, so *Hindutva* was made possible by the particular Nehruvian conception of socialism, the communal language implicit within the policies of affirmative action, and the specific electoral strategies of parties that stressed the importance of communities as aggregates of votes (Kaviraj 1995). Just as the Emergency was not inevitable, the rise of the BJP was contingent on a specific weakening of Nehru's improvised conception of secularism and citizenship, and the mobilisation and fragmentation of the caste–class alliances at the heart of the Congress system. The decline in Jaffrelot's key 'political variable' – the hegemonic secularism of the central state – occurs incidentally not through any emergence of an 'authentic' national identity, and not arguably through the mutual incomprehension between the elite and the subaltern (Corbridge and Harriss 2000; Kaviraj 1997), but because as a legacy of *garibi hatao*, the Emergency and the Janata period, the state leadership of Congress-I *abandoned* it, after failing to implement a coherent program of social emancipation.

Or rather, Indira Gandhi abandoned it on behalf of a wider elite, either as a conscious substitution for the strategy of *garibi hatao*, or more likely as a series of improvised shifts towards a discourse that would apparently co-opt dominant social forces that seemed to be collecting around an emergent, coalescent Hindu majority, which in the wake of the Emergency and the Janata government seemed particularly vulnerable both to the socio-economic assertion of the lower castes, especially the OBCs, and to the conspicuous displays of newly acquired wealth by Muslims.[1] As the Congress continued to disintegrate, and as caste and class mobilisations amplified the fragmentation of the Indian political system, *Hindutva* – and the BJP – came to the fore.

Yet the relationship between the BJP and *Hindutva* is itself a complex, dialetic relationship between an ideological strategy and an electoral search to acquire political office. From 1989 to 1996, the dialectics within the Sangh Parivar are not dissimilar to those which occurred between Gandhian socialists and Nehruvian socialists within the Congress, as the party approached national power in the mid- to late 1940s. Ultimately the Nehruvian emphasis on the state won out over an organicist, agrarian conception of society and local democracy. As Hansen states:

> Gandhi and the entire cultural nationalist tradition in India saw the essence of the nation as residing in India's cultural communities, whereas the political realm of the colonial state remained a morally empty space, a set of lifeless procedures and culturally alien institutions that could only be given life and indigenous meaning by a vibrant national community outside the political realm.
>
> (Hansen 1999: 50).

This view was inherited and advanced by the Hindu nationalists. Yet this distinction – of two worlds, one political, the other national/cultural, a societal world – is overdrawn, and because in the case of the BJP–VHP–RSS, it ignores the fact that the contradictions and synergy between and within these organisations takes place within the same space, a continuum of state–society relations, intertwining a whole myriad of institutions from Delhi down to the panchayat ray and municipal authorities. An emphasis on state–society relations shows that these two principles of organisation are intimately related, curled up into each other, being neither vertically or horizontally separable. From its political reincarnation in 1980 the BJP focused on the state, and the RSS focused on the long-term transformation of society, and together they created a constant tension between an aggregational electoral strategy (implying by 1996 coalition) and the sangathanist strategy favouring ideological realignment and Hindu purity throughout the federal system. But in practice, these two strategies were never as separate as conventional analysis portrays (Jaffrelot 2005; Noorani 2000).

They were – and are – constantly interacting. In specific conditions, these two routes to power could be reconciled relatively easily. In moments of crisis and communal polarisation, the 'impossible assimilation' yields electoral and ideological dividends. But unless an atmosphere of constant and sustained crisis could be maintained, the single issue of *Hindutva* is exposed as a narrow elitist project whose association with forward caste, middle class Indians is less contingent on its social transformation and spiritual renaissance than its promise of controlling the poor. In power, the BJP is constantly pressurised to provide and maintain policy areas in industry, agriculture, education and foreign policy, well away from the pristine minimalism of the shakars and the 'god' men, and under the constant scrutiny and active resistance of the poor through multiple institutional arenas (an assertive president, a growing and powerful Supreme Court, panchayat elections). With reference to activities of the sangh parivar, Barbara Harriss-White notes:

> India's is a society able to expose and contest such activity, of late notably through the Supreme Court, the National Commission of Human Rights, the activist media and many incidents of resistance, even though such freedom has not prevented the escalation of communal violence.
>
> (Harriss-White 2003: 250)

One of the main arguments offered here is that such contestation also involves the myriad institutions of the state itself as conduits for organising and empowering forces within society as counter-hegemony to *Hindutva*.

Given the diffuse nature of the social forces arrayed against *Hindutva* (both strategically as a project, and tactically as an elitist view of governance), the state is needed as a crucial organsing principle to articulate what are formidable levels of coordinating opposition through myriad state structures and a complex, tessellated society. This dilemma is of course shared to some extent by the BJP, but amid a fragmented opposition and based on a spatially consolidated constituency, it has been able to consolidate a large percentage of seats on a small popular vote. Through the fear of disorder, brought on by the decline of the once dominant Congress, the xenophobic discourses of Hindu nationalism developed in the heart of a large and expanding middle class.

This degree of consolidation was made possible by the tendency of OBC and backwards castes to compete within themselves and resist voting with scheduled castes and tribes, to isolate the BJP vote. This process was made easier for the BJP through the strategic use of violence by aspects of the Sangh Parivar that could, in very specific circumstances, polarise the vote and corral a Hindu votary. But these circumstances could not be sustained. As Uttar Pradesh was to demonstrate in 2004, where the BJP faced a combined caste-led opposition (linking the OBCs, backward castes and adivasis), the forward caste votes of the Hindu nationalists cannot win seats. In 2004, a surprise defeat for the NDA rested on a perception that the BJP was indeed pro-rich and a recognition by the Congress that coalition and power-sharing was the way to gain power and, through the state, to articulate not just a defence of secularism, but to attack the socio-economic record of the NDA government. Frankel observed that:

> The 2004 election campaign gradually created a consensus among secular parties not only that the BJP-led NDA must be prevented from returning to power, but at the national level the Congress still repre-sented the strongest bulwark against communalism
>
> (Frankel 2005: 778)

But it was also the recognition that the NDA's period in government was less impressive than the party itself claimed and was perceived as largely regressive by the Indian electorate at large.

Coalition government and the changes it has brought about throughout the whole web of state–society relations provide the basis for active resist-ance as well as compromise with *Hindutva*. The most visible – and perhaps the most superficial – change in governance has come about in the ways in which the Lok Sabha operates, requiring far more emphasis on cross-party support, and on what had been, under the Congress years (especially under Mrs Gandhi), a moribund and ineffectual committee structure. Governments devoid of large majorities cannot ignore the mood of the house and push through policies without debate. As the political complexion of the states has changed, the make-up of the Rajya Sabha has diverged from that of the Lok Sabha, and there have been numerous instances in which legislation has been returned from the upper house for reconsideration and amendment. Although ineffective in pressurising the national government to dismiss the Modi government, the Congress was able to utilise a series of procedures in 2003–2004 to get Gujarat debated. As state and national governments have fragmented, the electoral college for the Union presi-dent has required a more conscious effort to find – and elect – consensual candidates.

Coalition and minority governments have complicated the role of the Union presidency, but also given it more power. This is not just with ref-erence to calling and dismissing unstable coalition governments, but in advising prime ministers and increasingly referring national and state legislation back for reconsideration, or to the Supreme Court for a wider, informed view as to its implications (Hewitt and Rai forthcoming; Manor 1993). Indian presidents have been restored to the role imagined for them at the founding of the Republic, as set out in the constituent assembly debates, an elder statesmen (as yet there have been no women), a Republican analogy of the British monarchy, to advice, to caution and to warn. Rightly envisaged at the time of independence as a key relationship at the heart of the division of power within the constitution, the recent change has impacted on other institutions.

One concerns the use of Article 356. National coalitions, in which the Union governments rely on the support of their regional parties (in power in just one or two states) has restricted significantly the scope of President's Rule. In conjunction with the Supreme Court's *Bommai* judgment (1992, see above), the procedures and substantive claim for invoking Article 356 has fallen under judicial review. On this basis, Article 356 was refused on two occasions, in 1997 and 1998, and from 1999 until 2004 was only used a modest four times (Adeney 2005: 111). Indeed the only major controversy to erupt over Article 356 (in startling contrast to the mini-revolt during Mrs Gandhi's last term in office, or the long years of misuse from 1969 onwards) was the decision by the NDA to *not* use it against the BJP Modi government in Gujarat. The NDA's own constitutional review set up in 1999 as part of its manifesto reiterated the need to use Article 365 only in extremis but declined to abolish it.[2]

Other institutions have changed their shape and their modus operandi. Judicial activism had almost transformed the Supreme Court, not just with reference to the scope and influence of judicial review, but with reference to social activism (Austin 1993; Kothari 1993; Rao 2005). In the late 1980s, the Rajiv Gandhi government legislated to create India's first legal aid agency, following the Supreme Court's decision to allow third party representation with reference to public interest litigation. Combined with the growth in NGOs and voluntary organisations, there occurred overnight a dramatic change in the extent to which the Supreme Court was prepared to intervene into politics, now broadly defined, thereby reiterating a commitment to social justice and the minorities. The most dramatic intervention by the Supreme Court came in the wake of the Gujarat riots when it ordered both a re-trial of the *Best Bakery* case,[3] and furthermore, ordered that it took place outside of the state. In overturning the earlier verdict and in tacitly recognising that the judicial system in Gujarat might have been compromised to the point of no longer being able to defend minority rights against officially sanctioned persecution, the court was widely praised, although the failure of the earlier fast-track legal system was noted as well by human rights activists (BBC, 7 October 2006, <news.bbc.co.uk/1/hi/world/south_asia/4745926.stm>).

In contrast to the dialectics of the 1950s–1970s, in which courts were invariably cast in the role of defending property rights against an activist and reformist state, from the late 1980s onwards the roles appear to have been reversed. In championing civil rights, the courts may well be responding to a strategic withdrawal of the state, confronting the collapse of earlier state-led projects aimed at emancipating the poor. Property rights are no longer an issue of dispute with the government, because various national governments have long abandoned any serious plans of challenging them. Although the courts have been accused of supporting and legitimating the BJP's use of *Hindutva*, they have also in numerous rulings contested it. At the same time, national governments have increasingly resorted to the courts as a way of defusing potentially explosive situations. To some extent, such reliance upon the law has added to the narrow politicisation of the courts. It has also increased the exposure of the courts to the risks of defying large elected parties and being accused of anti-democratic behaviour.[4] Yet the court has in part responded by pushing forward new checks and balances to the Indian state as a whole.

Elections and new parties have fundamentally changed the nature of the MPs and the MLAs who represent their fellow Indians. The makeup of the Lok Sabha in turns of occupation has moved from an assembly dominated by lawyers to one dominated by agriculturalists and rich peasants, and slowly, to improve on the representation of the dalits and tribals. Yet despite an improvement in the general representative nature of government, there has come a massive increase in the number of criminals and 'goondas' being elected to parliaments, and general inaction as coalitions have been premised on minimum programs or sheer opportunism. Complex coalitions in the states – specifically Uttar Pradesh – wherein access to power provides the only incentive for otherwise contradictory alliances (such as a short-lived one between the BJP-BSP) oddly created a form of caste leader who 'ruled through authoritarian structures' and see their writs primarily in terms of caste politics (Frankel 2005: 722).

This has continued to degrade the ability of local and state governments to implement policies, or where this has been the case, frequent changes in government have led to U-turn politics or populist outbiddings through cheap or free provision of services. Liberalism at the national and state level has changed the strategies deployed by local elites to retain patronage within the regulatory structures overseeing economic reform. The proliferation of criminal activities through the state, and through the opportunities election to office presented, were equally present in municipal authorities, where paradoxically, the extension of elected authorities has created undemocratic outcomes and increases in corruption (Dahiya 2005). In such a light, substantive issues of democracy are sustained by mere procedures – often the mere act of voting.

There are two main objections to what is, in effect, an argument that *Hindutva* has contested a weakened and reactive state, but in the process of doing so, has been alchemised by the compulsion of the political and electoral process – in effect, it was compelled to separate out into antagonistic forces occupying discreet and multiple levels of state and society, in conflict over the price of retaining power. The first objection is the largely conventional argument that such a view of politics is too narrow to understand adequately the forces of Hindu nationalism. John Zavos put this view recently when he stated that:

> Formal politics and the control of the state is significant, but it needs to be placed within the context of broader forces, which conceptualise society as a range of segmented areas and 'functional groups', as Golwalker would have it. This point is graphically demonstrated by the network of organisations that constitute the Sangh Parivar. These organisations focus on a variety of issues, from tribal welfare, to education to labour relations, and this is an expanding network across areas of social and cultural life.
>
> (Zavos 2005: 51)

As such, the reductionism of politics to formal institutions (and formal institutions to the state) ignores the complex synergy between movement and party, governance and ideological forces, and pays disproportionate attention when these seem to be in conflict as opposed to when they are insidiously working side by side to change social, cultural and religious signifiers. It also disregards the evidence that, even through formal coalition with parties that do not share its ideological moorings, the societal and ideological agenda is advanced beyond the narrow class–caste base of the BJP, through social activities coordinated through the Sangh Parivar, such as health and educational provision in tribal areas (Adeney 2005: Jaffrelot 2005). As noted, considerable evidence exists to show that middle class opinion and its views on Muslims and more recently Christians have changed dramatically (Robinson; Anand 2007). And whatever comfort can be taken from the public exposure of M. M. Joshi's attempted educational coup, there is widespread evidence that, under the NDA, generic aspects of India's national education policy have been seriously compromised by the BJP's ideological links with the RSS (Lall 2005: 153–70).

The second objection is that the mainstreaming of the BJP through the logic of the political system ignores the pre-existing shortcomings and setbacks within this system itself, and while it may be praised for divorcing the BJP as a political force from the confrontational and potentially violent objectives of say the VHP and the RSS, the system is *still* moribund, or to paraphrase Kothari, still incapable of accommodating the poor. In power, the BJP has suddenly appeared to be both prone to corruption, factional disputes over the spoils to office, and including highly personalised state units that operate with overtly populist and authoritarian tendencies (Gurharpal Singh 2005; Varshney 2002). There have even been some suggestions that the tensions and exposures of power had impacted on the RSS itself and have, as in the wake of the NDA's defeat, generated a renewed dislike of politics, a greater interest in local institutions of power and belief that the wider world of state and national institutions as unclean and polluting. If, in response to the 2004 electoral setback, the BJP re-emerges as a Hindu nationalist party, shaped more on the old Congress of aggregation and co-optation without transforming and coming up with new forms of governance, it is equally as likely that it will end up encountering the same disjuncture between democracy and governance that was, in the first instance, the legacy of the Congress.

In dealing with the first point, while politics in India cannot be understood without taking culture and religion seriously as much neglected subjects in the general literature, they should not be reified or placed outside a political terrain that sees the state as 'above and apart' from society, and thus, in extremis, irrelevant to social forces. In using Rueschemeyer's emphasis on state–society relations, through Migdal, I have sought to bring into the remit of politics the intrinsic use of religious and cultural capital in identifying how state authority is actually constituted within society and influenced by social actors at multiple levels of the system itself. Writers such as van der Veer and Zavos tend, perhaps unintentionally, to reproduce two quite separate domains – the state and society – which then define two quite separate arenas in which the RSS and the BJP operate. This is misleading. For example, it was state policy that allowed the RSS to colonise various cultural fronts, and state failure to challenge it, that allowed the RSS to prosper, especially in the wake of the Emergency and through the Janata period, and has indeed defined the arena in which the RSS is most active – above all the neglect of India's educational infrastructure by the state (Chatterjee 2004: 113–30). Neglect by the state has left primary education and welfare in such an inadequate position that it is vulnerable to private initiatives and organisations favoured by the BJP (and the RSS) (Harriss-White 2003; Lall: 2005; Pandey 2006).

The second point returns the argument to the nature of Indian democracy and a position, derived in the main as a legacy of the deinstitutionalisation debates, that despite record levels of mobilisation, disproportionately affecting the poor, the party system and the wider institutions of the central state are incapable of accommodating mobilised political, subaltern demands in a systematic and accountable fashion. Rajni Kothari's work, which during the 1970s and 1980s so aptly described this growing disjuncture with reference to Mrs Gandhi and the Emergency, hoped that the rise of new social movements would provide the initiative required to restore the democratic deficit. The evidence of the 1990s is that these have tended not so much to redefine and reinstitutionalise politics as led to a form of privatisation in which the state further withdraws from a commitment to provide and administer social policy. This, a deliberate policy by the state, in turn reconfigures more societal arenas for the forces of *Hindutva* to become active, even as it generates resistance to policies through political opposition and the ability of political society to contest

Hindutva's hegemonic designs. The real need here seems to identify the class and caste basis of resistance and to move to coordinate *these* forces, but through the state, to give them coherence, that is, broadly accountable and able to implement policy.

This chapter has touched on areas where the onset of coalition has transformed the way in which specific institutions operate to improve their overall representative functions, and above else, to reconstitute checks and balances. Yet the biggest challenge here is the fragmentation of political authority that in part mirrors a fragmented social order premised on conflicting class–caste alliances. Utilising a term by M. N. Srinivas, Austin defined India in the late 1990s as a survival society, 'an orientation [that] produces an indifference to the well being of others and to the conditions of society as a whole, particularly on the part of those in the urban middle class' (Austin 1999: 641). *Hindutva* remains a threat because it holds out a simple and indigenous solution to this problem. But the state too can – and should – offer a solution to this, even if, as critics have so often pointed out, the central state and the constitution have been in part effectively co-opted by *Hindutva* itself.

It is one thing to challenge the *Hindutva* project, to expose its programmatic weakness, quite another to reinstitutionalise or reconfigure state–society relations towards an alternative coherent image of India that seeks to close the disjuncture between political mobilisation and democratic governance in less intolerant, less communal ways. Secularism in India, as it emerged through the fault line between affirmative action, group rights and private property, constituted the basis on which *Hindutva* was to be particularly successful, but it need not have done, as there was nothing intrinsic to this failure. It was a failure of political leadership – paradoxically perhaps by Nehru himself. Hansen rightly and succinctly remarked that 'in order to understand Hindu nationalism we need to analyse carefully the official secularism it opposed. We need to take a closer and a more informed look at the practice and meanings of secularism in the public culture of Independent India' (Hansen 1999: 11). But it does not follow from this (as Hansen would acknowledge) that there is only one form of secularism, or that the state cannot, through changing the basis of its uneven engagement with communities, seek to address the problem directly. In this sense, Bhargava is surly right when he notes that, in reiterating the desirability and inescapability of secularism, 'the real challenge before us continues to be one of working out an alternative conception of secularism rather than simply an alternative to it' (Bhargava 1998: 28). It is also necessary to look again at what socialism means and how historically the socialist project has been in India a form of elitism. There is not just a need to contextualise the word amid ongoing globalisation, but within a much more nuanced understanding of state and society and the processes and structures of regulation that determine policy and how it is – or is not – implemented.

Post-structural and anti-foundational scepticism over the state, especially the post-colonial state, have come close to condemning the state *per se* as the primary cause of authoritarian, hegemonic and violent strategies (Chatterjee 2004, Jalal 1995). I agree with Chatterjee's emphasis on the need to accommodate and co-opt innovative forms of governance from the *vox populi*, the multitude, what he calls political society but Chatterjee's thesis sees these incursions in too positive, too uncritical a light, and downplays the role played by the state – situated intimately at various levels within society, to crystallise both legitimate representation and legitimate denial (Midgal, Mann). Such critics of modernity have also appeared to validate societal resistance to the state for its own sake, and the development of parallel, para-legal authorities, along with the communities they represent, on the grounds that they contain innovative and more democratic forms of governance. This has come perilously close to condemning social violence.

What this book has set out to show is that although state co-optation of societal interests has been, in India, a narrow and instrumental affair, and has consistently exaggerated and politicised social movements to legitimate the specific claims of a given leadership, the state is not primarily – and certainly not necessarily – the locus of violent ideologies. Indeed, an alternative view can reveal that a weak state in India has too often moved to co-opt violent movements through a generous inclusion into the ambit of governmentality (Chatterjee 2004: 76). Citing works by Hansen and Nigam (reference), Chatterjee acknowledges the risks that political society spawns social violence as an end itself (to glorify, to redefine culture, to give a sense of belonging, and to further incidental political projects). Yet he has yet to include them into his analysis of political and civil society. States are not necessarily, and are certainly not generically, authoritarian institutions discharging 'toxins' of order and exclusion throughout multi-ethnic, pluralist societies. They may, with the right leaderships and the right strategies of mobilisation, constitute the only necessary antidote.[5] Understanding the constant interplay in which the state is both threat and protector requires a more sophisticated view of state–society relations than is usually found in post-structural and post-modern writings.

Notes

Introduction

1 One of the main arguments advanced here is that political theory often underplays random events and, in particular, unintentionality in analysing political outcomes. One of the advantages of Weberian sociology is it attempts to look at the unintended consequences that follow when political elites act out their roles within public institutions. Michael Mann rather interestingly offered to label such an explanation 'the state as "cock up"' (Mann 1993: 53). See Chapter 1.

2 This term is derived from P. Abrams, 'Notes on the difficulty of studying the state', *Journal of Historical Sociology*, I, 58–89. I would like to thank John Harriss for bringing this article to my attention.

1 Emergencies, states and societies: the study of Indian politics

1 I am referring here to a great deal of Marxist and neo-Marxist literature on state autonomy which, in seeking to separate politics from economics and cultural and social power from class, generated a vast quasi-apologetic literature that, with the onset of post-structural and post-modernist literature, seemed largely redundant.

2 The term 'kulak' is used extensively in debates on agrarian change in India. See Harriss (1982b) and Shanin (1987).

3 Political mobilisations 1963–1971

1 There are interesting parallels here in explanations offered for the deregulation and liberalisation strategies pursued with reference to industry in the 1990s – see Chapter 7.

2 In Shashi Tharoor's bitingly dry modern version of the Mahabharata, called (sensibly) *The Great India Novel*, Indira Gandhi is depicted as a solitary, brooding child much taken to playing with toy soldiers while her father is in prison (Tharoor 1989). More serious and less fictional accounts of her life dwell on her time spent separated from those she loved.

4 The state and political crisis 1971–1975

1 It is of interest that V. V. Giri initially rejected the advice and requested that the cabinet reconsider the question of Ray's appointment and that the president

expressed concern that the principle of seniority was being ignored. The decision was reconfirmed later the same day and the president gave way (Austin 1999: 279).

5 The state as political impasse 1975–1980

1 Bahuguna complained repeatedly about the interference of the PM's private assistant, Yashpal Kapoor, in state affairs. The joke at the time was that Bahuguna had been removed as the chief minister of Uttar Pradesh for 'interfering in the internal affairs of UP' (interview with S. S. Ray, 1986).

2 In an interview in 1986, S. S. Ray explained how, on hearing that a High Court judge had been transferred from West Bengal for ordering the release of Jyotirmoy Basu, he personally contacted the prime minister and has the order rescinded (interview with S. S. Ray, 1986).

6 Political mobilisations 1980–1986

1 The Congress split into Congress-I and Congress-U (Devraj Urs). It was argued that Indira Gandhi encouraged Charan Singh to defect from the Janata on the grounds that her supporters in the then Congress-I would support his claim to be prime minister. Following Charan Singh's defection and the fall of Morarji Desai's government, Jagjivan Ram also staked a claim to form a government. When President Sanjiva Reddy asked him to state which parties supported him, he allegedly replied: 'Where are parties?' (interview with Jagjivan Ram, 1986).

2 Through the 1990s, the Supreme Court would, on a number of occasions, rule that people did not have an unfettered right to conversion.

3 Jaffrelot also cites, as do many cultural commentators on Indian politics, the dramatic effects of the televised version of the Ramayana on Indian perceptions of religiosity, the impact of the Pope's visit to India in 1986 and the Rajiv Gandhi's decision to ban *The Satanic Verses* (Jaffrelot 2006).

4 The most recent committee – headed by Justice Nanavati – submitted its report to the Indian Parliament in February 2005.

5 Again the dynamics here point to a problem with the uncritical approach Indian secularism had adopted, in practice, to what constituted religious 'communities' (see Chatterjee 2004: 113–30).

6 See the Report of the Archaeological Findings at Ayodhya 2002–2003 (Archaeological Survey of India, New Delhi, 2003). The findings of the report have been subject to intensive criticism.

7 *Hindutva* as crisis 1986–2004

1 There is not enough space here to explore this idea further, but at one level the RSS-VHP conceptions of society surely require comparison with other religious militant organisations like Hamas and Ja'amt and the similar difficulties these organisations or movements face when they try to implement policy or rather confront policies that do not fit or are not adequately dealt with by their existing doctrines.

2 Ethnographic and cultural studies – post-structural and indeed post-modern in nature – have proved extremely useful in looking at the invention of processions, although few if any have begged what might be the interesting associations between these and fascism.

3 This was later converted into an overall majority when the Congress-I absorbed an entire party – the Jharkhand Mukti Morcha.

4 See the interesting discussion in Crispin Bates over how the consolidation of lower caste reservations within the state might well have encouraged members of the intermediate classes to see their future in integrating India into global markets, and not on resources and subsidies channelled to them through the state, which they no longer had the power to exclusively obtain (Bates and Basu 2005: xv).

5 In the wake of the V. P. Singh government's decision over Mandal, the BJP leadership had been careful to protest over the use of caste–jati designations for reservations, and not the principle of reservation – or affirmative action – in particular.

6 In Uttar Pradesh, in 1993, the BJP secured some OBC votes, but this seems to have been the result of inter-caste rilvary between Lodhis and Kurmis vis-à-vis the Yadavs.

7 Jaffrelot contests the ease with which the press – and also specific commentators – use the terms moderates and extremists, since it conveys a misleading impression over the relationship between the BJP and the RSS.

8 See 'RSS Unhappy', *The Hindu*, 2 October 1998. Frankel's chapter on the 'Challenge of Hindu Nationalism' outlines how pressures and disagreements between the BJP and the RSS were actually projected into cabinet meetings.

Conclusions

1 This remains a significantly under-researched area – the so-called wealth of the Muslim community was more often asserted than researched in the 1980s.

2 An interesting comparison can be made between the findings of the National Commission to Review the Workings of the Consitution (NCRWC) and the earlier submissions by non-Congress parties to the Sarkaria Commission in the early 1980s. A long stint in power has clearly modified their views to what were widely held earlier to be 'authoritarian and unacceptable aspects' of the constitution.

3 In 2002 14 people, 12 of them muslims, were burned to death in a bakery shop in Vadordara. In 2003, 21 Hindus accused of the murder were acquitted by a fast-track legal system set up to deal with the crisis in Gujarat, due to lack of evidence when witnesses changed their testimony. It was alleged that prominent BJP politicians had threatened the prosecution witnesses. Amid criticism of police failure, the case was retried in nearby Maharashtra.

4 One explanation put to me over the Supreme Court's ruling on *Hindutva* was that the court had to accept the electoral presence within the Lok Sabha of Hindu Nationalists that, by 1993, made up the second largest party. In such circumstances, it could not risk unseating them by upholding the claim that, for the purposes of the Representation of the Peoples Act, *Hindutva* was an incitement to religious hatred: 'The law is not a substitute for a failure of political will.' Or, indeed one might add, for politics.

5 It is tempting to conclude, in a final comparison with the Emergency, that the threat of Indian fascism – as clearly manifested by *Hindutva* – derives not from the state at all but from social movements themselves and their co-optation by the weak state. Even if one takes the largely European debates on fascism that stress state–society relations and a crisis brought about by dominant class factions unable to assert their control through democracy and thereby ensure

continued capital accumulation, aspects of the *Hindutva* ideology are suggestive of fascist social civil 'organisations'. Through 'crisis' – not necessarily defined through economics alone – such forces perceive their chance to take over the state as the culmination of its strategy. *Hindutva's* ability to impose a corporatist, organicist society and then mould the state accordingly is, as has been argued here, unlikely to succeed unless the state is compelled to cooperate. Even then, contradictions within the project itself remain.

References

Government Documents

Government of India.
Constituent Assembly Debates (CAD)
Volumes 1–12
(19 December 1946–26 November 1949)

Constituent Assembly
Constitutional Precedents (First Edition)
(2 September 1946).

Lok Sabha Debates (LSD)
Fourth Series Volumes 1–47
(16 March 1967–18 November 1970)

Fifth Series Volumes 1–65
(19 March 1971–5 November 1976)

Sixth Series, only one Volume (25 March 1977)

Government of India. Home Ministry Reports:
1967–1968
1968–1969
1969–1970
1970–1971
1971–1972
1972–1973
1973–1974
1974–1975
1975–1976
1976–1977
(Obtained from Indian Institute of Public Affairs, IIPA, New Delhi.)

Government of India. Administrative Reforms Committee Report. Report on the Machinery of Government of India and its procedure of Work. 16 September 1968.

Government of India. Administrative Reforms Committee Report. Report on Centre–State Relations. 19 June 1969.

Government of India. (Home Ministry). 'Why Emergency?,' 7 July 1975.

Government of India. (Information and Broadcasting Ministry). 'Misuse of the Mass Media', 1977.

Government of India. (Matthew Commission of Enquiry). Enquiry into the Explosions that took place in Samastapur, Bihar 1975, New Delhi October 1975

Government of India. (Shah Commission of Enquiry) Enquiry into the Emergency. Interim Reports I, II and Final Report. New Delhi, 11 March 1978.

Government of India. (Jagmohan Reddy Commission of Enquiry) Enquiry Regarding the Nagarwala Case. Final Report. New Delhi, 23 October 1978.

Government of India. Commission of Enquiry into the Maruti Affair. Final Report. New Delhi, 31 May 1979

Government of India. Report of the Commissioner on Scheduled Castes and Scheduled Tribes. 24th Report 1975–1977

Government of India. (Electoral Commission of India) Report on the Fourth General Elections in India. Vol. I General. Vol. II Statistical.

Government of India. (Electoral Commission of India) The Fourth General Elections: An Analysis.

Government of India. (Electoral Commission of India). Report on the Fifth General Elections in India. Volume I: General. Volume II: Statistical.

Government of India. (Electoral Commission of India). Report on the Mid-Term Poll results in Uttar Pradesh, Orissa (1974) and Gujarat (1975)

Government of India. (Electoral Commission of India). Report on the Sixth Lok Sabha Elections 1977. Report on the Legislative Assembly Elections 1977–1978.

Government of India. (Electoral Commission of India). Report on the Seventh General Elections to the Lok Sabha 1980. Report on the Legislative Assembly Results 1980–1981.

Party Documents

Indian National Congress

Details from the minutes of All India Congress Committee (AICC), Party Plenary Sessions, Meetings of the Congress Working Committee(CWC), Annual Reports from the General Secretaries (taken from:

Zaidi, A. M., *Encyclopaedia of the National Congress*, New Delhi, 1983.

Vol. 19, 1966–1967

Vol. 20, 1968–1969

Vol. 21, 1970–1971

Vol. 22, 1972–1973

Vol. 23, 1974–1975

Vol. 24, 1976–1977

Zaidi, A. M., *Annual Register of Political Parties in India* (1972–1973).

Zaidi, A. M., *The Great Upheaval. Party Documents relevant for the Split* (1969–1972).
Congress-O Pamphlet, 'The Congress Split', by A. Ghosh (with an introduction by Nijalingappa).
Indian National Congress: Party Report on the Fourth General Elections. Janata Party Publication: 'Development For Whom?', 1979.
Open letter to All Congressmen: Correspondence between H. N. Hahuguna and Prime Minister Indira Gandhi, 1980.
Janata Party Publication: Report on Centre–State Relations, 1983.
Lok Dal Party: Working Paper on Centre–State Relations, 1980.

Interviews

P. N. Haksar, Principal Private Secretary to the Prime Minister (1969–1973), Minister of State for Planning 1976–1977. Interviewed 19th April 1985.
V. C. Shukla, Minister of State for Planning 1974–1975, Minister of State for Information and Broadcasting 1976–1977. Interviewed 20th April 1985.
I. K. Gujral, Minister of State for Information and Broadcasting 1974–1975, Minister of State for Planning 1976 (up to 12.5.76). Interviewed 10th May 1985.
D. K. Barooah, Cabinet Minister for Petroleum and Chemicals (1974–1975), President of the Indian National Congress 1975–1977. Interviewed 5th May 1985.
S. S. Ray, Chief Minister of West Bengal 1972–1977 (previously Minister of State for Education and Welfare). Interviewed 6th February 1986.
Om Mehta, Minister of State for Home Affairs, Personnel and Administrative Reforms and Parliamentary Affairs (1975–1976). Interviewed 19th February 1986.
H. N. Bahuguna, Minister of State for Communications (1971–1975), Chief Minister of Uttar Pradesh 1975. Resigned in January 1977 to co-found Congress for Democracy (CFD). Interviewed 26th February and 2nd March 1986.
Mohan Dharia, Minister of State for Planning, Works and Housing (1975) (previously Minister of State for Planning). Interviewed 13th February 1986.
Dinesh Singh, Minister of External Affairs (1969), Minister for Industrial Development and Internal Trade (1970). Interviewed 21st February 1986.
S. Subramaniam, Senior Civil Servant Home Ministry, Fellow of Nehru Memorial Library, New Delhi, 1985–1988. Interviewed 10th February 1986.
P. Sen Gupta, Research Professor at Centre for Policy Studies, New Delhi. Interviewed 4th June 1985.
Mark Tully, BBC correspondent in India since 1966. Interviewed 2nd May 1985.
Aswini Ray, Professor at the Centre for Political Studies, JNU, New Delhi. Interviewed 30th April 1985, 17th February 1986.
Rajni Kothari, Professor at the Centre for the Study of Developing Societies. Interviewed 5th June 1985.
Madu Dandavate, Professor at JNU, Minister of Railways/Minister of Commerce and Civil Supplies (1977–1979), Leader of the Janata Party in Parliament 1985–1986. Interviewed 23rd February 1986.
Maneka Gandhi, Widow of Sanjay Gandhi, Founder of the Sanjay Gandhi Forum. Interviewed 25th February 1986.

Francine Frankel, Professor at University of Pennsylvania. Interviewed at the International Centre, New Delhi, 30th April 1985.
The following declined to be interviewed:
Bansi Lal
Abdul Ghafoor
R. K. Dhawan

Newspapers and Magazines

(Indian Newspapers taken from the Newspaper Archives at Sapru House Library, New Delhi, JNU Library, South Delhi, and the Nehru Memorial Library, New Delhi.)

- *The Statesman* (Calcutta and New Delhi)
- *The Statesman Weekly*
- *The Hindu* (Madras)
- *The Times of India* (Bombay and New Delhi)
- *The Indian Express* (New Delhi)
- Sunday Standard
- *Hindustan Times*
- *The Motherland*
- *The Patriot*
- *The National Herald* (Lucknow)
- *The Assam Tribune*
- *Free Press Journal* (Bombay)
- *The Tribune*
- *Amrita Bazar Patrika* (Calcutta)
- *Northern India Patrika*
- *The Economic Times*
- *The Guardian* (London)
- *The Times* (London)
- *The International Herald Tribune* (London)
- *Seminar*
- *The Economist*
- *The Illustrated Weekly of India*
- *Economic and Political Weekly*
- *The Eastern Economist*
- *Quarterly Economic Report* (IIPA, New Delhi)
- *Monthly Commentary on Indian Economic Conditions* (IIPA)

Secondary Sources

Abrams, P. (1988) 'Notes on the difficulty of studying the state', *Journal of Historical Sociology*, 1: 58–89.
Adams, J. (1990) 'Breaking away: India's economy vaults into the 1990s' in M. Bouton and P. Oldenburg (eds), *India Briefing 1990*, Boulder, CO: Westview Press, 77–100.

Adeney, K. (2005) 'Hindu Nationalists and federal structures in an era of regionalism' in K. Adeney and L. Saez (ed.) *Coalition Politics and Hindu Nationalism*, Oxford, Routledge.

Adeney, K. and Saez, L. (eds) (2005) *Coalition Politics, and Hindu Nationalism*. London: Routledge.

Ahmed, B. (1970) 'Caste and electoral politics', *Asian Survey*, 10: 979–93.

Ahluwalia, I. J. (1995) *Industrial Growth in India: Stagnation Since the Mid-Sixties*. New Delhi: Oxford University Press.

Alavi, H. (1973) 'State in post colonial society', in R.Gough and H. P. Sharma (eds), *Imperialism and Revolution in South Asia*, New York, 291–337.

Anand, D. (2007) 'Anxious sexualities: masculinity, nationalism and violence', *British Journal of Politics and International Relations*, 9: 257–69.

Asher, C. (1992) *Architecture of Mughal India*. Cambridge: Cambridge University Press.

Austin, G. (1999) *Working a Democratic Constitution: The Indian Experience*. New Delhi: Oxford University Press.

Balasubramanyam, V. N. (1984) *The Economy of India*. London: Weidenfeld and Nicolson.

Banerjee B. Nath. (1982) *Religious Conversions in India*. New Delhi: Harnam Publishing.

Bardhan, P. K. (1978) 'Authoritarianism and democracy', *Economic and Political Weekly*, 13: 529–33.

Bardhan, P. K. (1984) *The Political Economy of Development in India*. Oxford: Blackwell.

Barnett, R. (1975) "Economic liberalisaton and the Indian state', *Third World Quarterly*, 7: 196–201.

Barrington Moore (1993) *Social Origins of Dictatorship and Democracy: Lord and Peasant in the Making of the Modern World*. Boston: Beacon Press.

Basu D. D. (1961) *Commentary on the Indian Constitution*. New Delhi: Oxford University Press, 125–50.

Basu, S. (2005) 'Political institutions, strategies of governance and forms of resistance in small market towns of contemporary Bengal: a study of Bolpur municipality', in C. Bates and S. Basu (eds), *Rethinking Indian Political Institutions*, London: Anthem Press.

Bates, C. and Basu, S. (eds) (2005) *Rethinking Indian Political Institutions*. London: Anthem Press.

Bayley, D. H. (1962) *Preventive Detention in India*. Calcutta: Oxford University Press.

Bayley, D. H. (1969) *The Police and Political Development in India*. Princeton: Princeton University Press.

Bayly, S. (1999) *Caste, Society and Politics in India from the Eighteenth Century to the Modern Age*. Cambridge: Cambridge University Press.

Bekerlegge, G. (2004) 'The Rashtriya Swayamsevak Sangh's "tradition of selfless service"', in J. Zavos *et al.* (eds), *The Politics of Cultural Mobilization in India*, New Delhi: OUP, 105–35.

Bhargava, R. (1998) *Secularism and its Critics*. New Delhi. Oxford University Press.

Blair (1977) 'Mrs. Gandhi's emerging and the Indian elections of 1977', *Modern Asian Studies*, 14: 237–70.

Block, F. (1980) 'Beyond relative autonomy: state managers as historical subjects', *The Socialist Register*, 227–42.

Bose, S. and Jalal, A. (1998) *Modern South Asia: History, Culture, Political Economy.* New Delhi: Oxford University Press.

Brass, P. R. (1965) *Factional Politics in an Indian State: The Congress Party in Uttar Pradesh.* Berkeley: University of California Press.

Brass, P. R. (1984) 'National power and local politics in India: a twenty year perspective', *Modern Asian Studies*, 18: 89–118.

Brass, P. R. (1989) 'Pluralism, regionalism, and decentralising tendencies in contemporary Indian politics', in A. J. Wilson and D. Dalton (eds), The *State of South Asia: Problems of National Integration*, London: Hurst Publishing, 223–64.

Brass, P. R. (1991) *Ethnicity and Nationalism: Theory and Comparison.* New Delhi: Sage.

Brass, P. R. (1994) *The Politics of India Since Independence*, 2nd edn. Cambridge: Cambridge University Press.

Brass, P. R. (1997) *Theft of an Idol: Text and Context in the Representation of Collective since Violence.* Princeton: Princeton University Press.

Brass, P. R. and Vanaik, A. (eds) (2002) *Competing Nationalisms in South Asia: Essays for Asghar Ali Engineer.* Hyderabad: Orient Longman.

Brecher, M. (1966) *Succession in India: A Study in Decision-making.* London: Oxford University Press.

Brecher, M. (1969) *Political Leadership in India.* London: Oxford University Press.

Brown, J. M. (1985) *Modern India: The Origins of an Asian Democracy.* New Delhi: University Press.

Brown, J. M. (2003) *Nehru: A Political Life.* New Haven: Yale University Press.

Burnell P. (ed.) (2006) *Democratization Through the Looking-glass.* Manchester: Manchester University Press.

Burnell, P. and Ware, A. (eds) (1998) *Funding Democratisation.* Manchester: Manchester University Press.

Butler, D., Lahiri, A. and Roy, P. (1984) *Compendium of Indian Elections.* New Delhi: Arnold-Heinamen.

Byres, T. (ed.) (1997) *The State, Development Planning and Liberalisation in India.* New Delhi: Oxford University Press.

Byres, T. (ed.) (1998) *The Indian Economy: Major Debates Since Independence.* New Delhi: Oxford University Press.

Chatterjee, P. (2004) *The Politics of the Governed Reflections on Popular Politics in Most of the World.* New York: Columbia University Press.

Chatterjee, P. and Mallik, A. (1997) 'Indian democracy and bourgeois reaction', in P. Chatterjee (ed.), *A Possible India: Essays in Political Criticism*, New Delhi: Oxford University Press, 35–42.

Chiriyankandath, J and Wyatt, A. (2005) 'The NDA and Indian foreign policy', in K, Adeney and L. Saez (eds), *Coalition Politics and Hindu Nationalism.* London: Routledge, 193–211.

Corbridge, S. and Harriss, J. (2000) *Reinventing India: Liberalization, Hindu Nationalism and Popular Democracy.* Cambridge: Polity Press.

Das, V., Gupta, D. and Uberoi, P. (1999) *Tradition, Pluralism and Identity: In Honour of T. N. Madan* New Delhi: Sage.

Desai, A. R. (ed.) (1986) *Agrarian Struggles in India After Independence*. New Delhi: Oxford University Press.

Desai, M. (2005) *Development and Nationhood: Essays in the Political Economy of South Asia*. New Delhi: Oxford University Press.

Dhar, P. N. (2000) *Indira Gandhi, The Emergency and Indian Democracy*. New Delhi: Oxford University Press.

Dieckhoff, A. *et al.* (eds) (2005) *Revisiting Nationalism: Theories and Processes*. London: Hurst and Company.

Dumont, L. (1970) *Religion, Politics and History in India*. London: Oxford University Press.

Eat-well, R. (1996) *Fascism: A History*. London: Vintage Press.

Engineer, A. (ed.) (1984) *Communal Riots in Post-Independence India*. Hyderabad: Sangam.

Engineer, A. and Shakir, M. (eds) (1985) *Communalism in India*. New Delhi: Ajanta Books International.

Franda, M. (1962) 'The organisational development of the India Congress Party', *Pacific Affairs*, 35: 248–60.

Frank, K. (2001) *Indira: The Life of Indira Nehru Gandhi*. London: Harper Collins.

Frankel, F. R. (1971) *India's Green Revolution: Economic Gains and Political Costs*. Princeton: Princeton University Press.

Frankel, F. R. (1978) *India's Political Economy, 1947–1977: The Gradual Revolution*. Princeton: Princeton University Press.

Frankel, F. R. (2005) *India's Political Economy, 1947–2004: The Gradual Revolution*, 2nd edn. New Delhi: Oxford University Press.

Frankel, F. R. and Rao, M. S. A. (eds) (1989) *Dominance and State Power in Modern India: Decline of a Social Order*,Vol. 1. New Delhi: Oxford University Press.

Frankel, F. R. and Rao, M.S.A. (eds) (1990) *Dominance and State Power in Modern India: Decline of a Social Order*,Vol. 2. New Delhi: Oxford University Press.

Gadgil, D. R. (1961) *Planning and Economic Policy in India*. New Delhi: Oxford University Press.

Galanter, M. (1984) *Competing Equalities: Law and the Backward Classes in India*. New Delhi: Oxford University Press.

Galanter, M. (ed.) (1992) *Law and Society in Modern India*. New Delhi: Oxford University Press.

Galanter, M. (1998) 'Secularism East and West', in R. Bhargava (ed.), *Secularism and Its Critics*. New Delhi: Oxford University Press, 234–67.

Gallagher, J., Johnson, G. and Seal, A. (eds) (1973) *Locality, Province and Nation: Essays on Indian Politics, 1870 to 1940*. Cambridge: Cambridge University Press.

Gangal, S. (1972) *The Prime Minister and the Indian Cabinet*. New Delhi: Ajanta Books.

Ghosh, S. (1970) *The Disinherited State: A Study of West Bengal 1967–1970*. Bombay: Orient Longman.

Gopal, S. (1993) *Anatomy of a Confrontation: The Rise of Communal Politics in India*. London: Zed Books.

Gough, K. and Sharma, H. P. (eds) (1973) *Imperialism and Revolution in South Asia*. New York: Monthly Review Press.

Gould (2003) in P. Burnell (ed) (2006) *Democratization Through the Looking-glass* Manchester: Manchester University Press, 23–40.

Graham, B. (1990) *Hindu Nationalism and Indian Politics: The Origins and Development of the Bharatiya Jana Sangh.* Cambridge: Cambridge University Press.

Grewal, J. S. (1990) *The Sikhs of the Punjab.* Cambridge: Cambridge University Press.

Guha, R. (1998) *Dominance Without Hegemony: History and Power in Colonial India.* New Delhi: Oxford University Press.

Gupta, S. (1990) 'The gathering storm', in M. Bouton and P. Oldenburg (eds), *India Briefing 1990.* Boulder, CO: Westview Press, 25–50.

Haksar, P. N. (1979) *Premonitions.* Bombay: Interpress.

Haksar, P. N. (1990) *One More Life.* New Delhi: Oxford University Press.

Hansen, T. B. (1999) *The Saffron Wave: Democracy and Hindu Nationalism in Modern India.* Princeton: Princeton University Press.

Hansen, T. B. (2001) *Wages of Violence: Naming and Identity in Postcolonial Bombay.* Princeton: Princeton University Press.

Hansen, T. B. (2004) 'Politics as permanent performance', in J. Zavos *et al.* (eds), *The Politics of Cultural Mobilization in India.* New Delhi: OUP, 19–36.

Hansen, T. B. and Jaffrelot, C. (eds) (2001) *The BJP and the Compulsions of Politics in India,* 2nd edn. New Delhi: Oxford University Press.

Hanson, A. H. (1966) *The Process of Planning: A Study of India's Five-Year Plans 1950–1964.* London: Oxford University Press.

Hardgrave, R. (1970) 'Congress in India: crisis and split', *Asian Survey,* 10: 256–62.

Hardgrave, R. L. and Kochanek, S. A. (eds) (2000) *India: Government and Politics in a Developing Nation,* 6th edn. Fort Worth: Harcourt Brace.

Harrison, S. (1960) *India, the Most Dangerous Decades.* Princeton: Princeton University Press.

Harriss, J. (1982a) *Capitalism and Peasant Farming: Agrarian Structure and Ideology in Northern Tamil Nadu.* Bombay: Oxford University Press.

Harriss, J. (ed.) (1982b) *Rural Development: Theories of Peasant Economy and Agrarian Change.* London: Hutchinson.

Harriss, J. *et al.* (2004) *Politicising Democracy: The New Local Politics and Democratisation.* Basingstoke: Palgrave-Macmillan.

Harriss-White, B. (2003) *India Working: Essays on Society and Economy.* Cambridge: Cambridge University Press

Hart, H. (ed.) (1976) *Indira Gandhi: India's Political System Reappraised.* Princeton: Princeton University Press.

Heginbotham (1971, December) 'The 1971 revolution in Indian voting behaviour', in *Asian Survey,* 11(12): 1133–52.

Heston, A. (1990) 'Poverty in India: some recent policies', in M. Bouton and P. Oldenburg (eds), *India Briefing 1990.* Boulder, CO: Westview Press, 101–30.

Hewitt, V. M. (1989) *Ethnicity, Sub-Nationalism and Federalism in South Asia.* Southall, Middx.: Shakti Communications for Hull University.

Hewitt, V. M. (1994) 'Prime minister and the president', in J. Manor (ed.), *Nehru to the Nineties: The Changing Office of Prime Minister in India.* London: Hurst and Company, 48–73.

Hewitt, V. M. (1995) *Reclaiming the Past?: The Search for Political and Cultural Unity in Contemporary Jammu and Kashmir.* London: Portland Books.

Hewitt, V.M. (2000) 'Containing Shiva?: India, non-proliferation, and the comprehensive test ban', *Contemporary South Asia,* 9: 25–39.

Hinnells, J. R. and King, R. (eds) (2006) *Religion and Violence in South Asia: Theory and Practice.* London: Routledge.

Hiro, D. (1977) *Inside India Today.* London: Routledge and Kegan Paul.

Jaffrelot, C. (1996) *The Hindu Nationalist Movement and Indian Politics: 1925 to the 1990s: Strategies of Identity-Building, Implantation and Mobilisation.* New Delhi: Viking.

Jaffrelot, C. (2003) *India's Silent Revolution: The Rise of the Low Castes in North Indian Politics.* New Delhi: Permanent Black.

Jalal, A. (1995) *Democracy and Authoritarianism in South Asia: A Comparative and Historical Perspective.* Cambridge: Cambridge University Press.

Jannuzi, F. T. (1974) *Agrarian Crisis in India: The Case of Bihar,* Austin: University of Texas.

Jenkins, R. (1999) *Democratic Politics and Economic Reform in India.* Cambridge: Cambridge University Press.

Jenkins, R. (2005) 'The NDA and the politics of economic reform', in Adeney K. Adeney and L. Saez (eds), *Coalition Politics and Hindu Nationalism.* Routledge, London: Routledge, 173–92.

Jhabvala, G. (1980) *The Indian Constitution.* New Delhi: Janata Books.

Jones, K. W. (1989) *Socio-Religious Reform Movements in British India.* Cambridge: Cambridge University Press.

Joshi, V. J. (1989) 'Fiscal stabilization and economic reform in India', in Ahluwalia and I. Little (eds), *India's Economic Reforms and Development: Essays for Manmohan Singh.* New Delhi: Oxford University Press, 147–68.

Joshi, V. J. and Little, I. (1994) *India: Macroeconomics and Political Economy 1964–1991.* Washington, D.C.: World Bank Publications.

Kamal K. L. (1982) *Democratic Politics in India.* New Delhi: Sangam Books.

Kanungo, P. (2002) *RSS's Tryst with Politics: From Hedgewar to Sudarshan.* New Delhi: Manohar Books.

Kaur, T. O. (2005) *Sikh Identity: An Exploration of Groups Among Sikhs.* Aldershot: Ashgate.

Kaviraj, S. (1986) 'Indira Gandhi and Indian politics', *Economic and Political Weekly,* 21: 1699–1702.

Kaviraj, S. (ed.) (1997) *Politics in India.* New Delhi: Oxford University Press.

Khilnani, S. (1997) *The Idea of India.* London: Penguin.

Kochanek, S. (1968) *Congress Party in India: The Dynamics of One Party Democracy.* Princeton: Princeton University Press.

Kohli, A. (1987) *The State and Poverty in India: The Politics of Reform.* Cambridge: Cambridge University Press.

Kohli, A. (1988) *India's Democracy: An Analysis of Changing State-Society Relations.* Princeton, NJ: Princeton University Press.

Kohli, A. (1990) *Democracy and Discontent: India's Growing Crisis of Governability.* Cambridge: Cambridge University Press.

Kohli, A. (ed.) (2001) *The Success of India's Democracy.* Cambridge: Cambridge University Press.

Kohli, A. (2004) *State-Directed Development: Political Power and Industrialization in the Global Periphery.* Cambridge: Cambridge University Press.

Kohli, A. B. (1983) *The Council of Ministers in India 1947–1982.* New Delhi: Janata Books.

Kothari, R. (1970) *Politics in India.* New Delhi: Orient Longman.

Kothari, R. (ed.) (1970) *Caste in Indian politics.* New Delhi: Orient Longman.

Kothari, R. (1976) *Democratic Polity and Social Change in India: Crisis and Opportunities.* Bombay: Allied Publishers.

Kothari, R. (1977) 'Retrospect and prospect', in *Seminar,* New Delhi, March 2, 32–42.

Kothari, R. (1983a) 'The politicians', in I. Khan (ed.) *Fresh Perspectives on India and Pakistan*. Bougainvillea Books, Oxford, 196–201.

Kothari, R. (1983b) 'Representative institutions under threat', in J. Manor and P. Lyon (eds), *Transfer and Transfomation: Political Institutions in the New Commonwealth*. Leicester: Leciester University Press, 31–45.

Kothari, R. (1988) *Integration and Exclusion in Indian*. Southall: Shakti Communications for the University of Hull.

Kothari, R. (1994) 'Rise of the Dalits and the renewed debate on caste', in *Economic and Political Weekly*, 29: 1589–94.

Kothari, R. (1995) *Poverty: Human Consciousness and the Amnesia of Development*. London: Zed Books.

Kotz, D. M., McDonough, M. T. and Reich, M. (eds) (1994) *Social Structures of Accumulation: The Political Economy of Growth and Crisis*. Cambridge: Cambridge University Press.

Krishna, G. (1966) 'Development of the India National Congress Party as a mass organisation', *Journal of Asian Studies*, 25: 413–40.

Krishnan, T. V. K. (1971) *Chavan and The Troubled Decade*. Bombay: Hermat Books.

Kumar, M. (1982) *Social Equality: The Constitutional Experiment in India*. New Delhi: Sangam.

Kurian, K. M. and Varughese P. N. (1981) *Centre-State Relation*. New Delhi: Ajanta Books.

Kurian, M. V. (1986) *The Caste-Class Formations: A Case Study of Kerala*. New Delhi: B R Publishing.

Lall, M. (2005) 'Indian education policy under the NDA government', in K. Adeney and L. Saez (eds), *Coalition Politics and Hindu Nationalism*. London: Routledge, 153–70.

Lanzendoffer, M and Ziemann, W. (1977) 'The state in peripheral societies', in Miliband, R. and John Saville (eds) *The Socialist Register*, London: Merlin Press, 143–77.

Larson, G. J. (1995) *India's Agony over Religion*. New York: State University of New York Press.

Lele, J. (1979) *Elite Pluralism and Class Rule: Political Development in Maharashtra, India*. Toronto, Buffalo: University of Toronto Press.

Leys, C. (1989) 'The overdeveloped post-colonial state: a re-evaluation', in *Review of African Political Economy*, 5: 39–49.

Lieten, G. K. (1996) *Development, Devolution and Democracy: Village Discourse in West*. Thousand Oaks, CA: Sage Publications.

Lipton, M. (1995) 'Liberalisation, poverty reduction and agricultural development in India' in R. Cassen and V. J. Joshi (eds) *India: The Future of Economic Reform*. New Delhi: Oxford University Press, 327–38.

Low, D. A. (ed.) (1977) *Congress and the Raj: Facets of the Indian Struggle, 1917–47*. London: Heinemann.

Low, D. A. (1988) (ed.) *The Indian National Congress: Centenary Hindsights*. New Delhi: Oxford University Press.

Ludden, D. (ed.) (1996) *Contesting the Nation: Religion, Community, and the Politics of Democracy in India*. Philadelphia: University of Pennsylvania Press.

McLennan, G. (1981) *Marxism and the Methodologies of History*. London: New Left Books.

McLennan, G. (2006) *Sociological Cultural Studies: Reflexivity and Positivity in the Human Sciences*. Basingstoke: Palgrave Macmillan.

McMillan, A. (2005) 'The BJP coalition: partnership and power-sharing in government' in K. Adeney and L. Saez (eds), *Coalition Politics and Hindu Nationalism*. Routledge: London, 13–35.

Madan, T. N. (1987) *Non-renunciation: Themes and Interpretations of Hindu Culture*. New Delhi: Oxford University Press.

Madan, T. N. (1998) 'Secularism in its Place', in R. Bhargava (ed.), *Secularism and Its Critics*. New Delhi, Oxford University Press, 297–320.

Mann, M. (1993). *The Sources of Social Power: The Rise of Classes and Nation-States, 1760–1914*. Vol. 2. Cambridge: Cambridge University Press.

Mann, M. (2005) *The Dark Side of Democracy: Explaining, Ethnic Cleansing*. Cambridge: Cambridge University Press.

Manor, J. (1978) 'Indira and after', *Round Table*, 68: 318–32.

Manor, J. (1978) 'Where Congress survived', *Asian Survey*, 18: 785–803.

Manor, J. (1980) 'Party decay and political crisis', in *The Washington Quarterly*, 11 (Summer): 25–40

Manor, J. (1983) *Transfer and Transformation: Political Institutions in the New Commonwealth: Essays in Honour of W. H. Morris-Jones* (Peter Lyon and James Manor, eds). Leicester: Leicester University Press.

Marx, K. (1977) 'The class struggle in France', in *Selected Works*, Vol. 1. London: Lawrence and Wishart, 641–58.

Marx, K. (1977) 'The eighteenth Brumaire of Louis Bonaparte', in *Selected Works*,Vol. 1. London: Lawrence and Wishart, 96–179.

Marx, K. and Engels. F. (1977) 'Manifesto of the communist party', in *Selected Works*, Vol 1. London: Lawrence and Wishart, 35–63.

Mason, T. (1968) 'The primacy of politics: politics and economics in national socialist Germany', in S. J. Woolf (ed), *The Nature of Fascism*. London: Routledge, 117–132.

Mayer (1984) 'Congress (I) emergency (I): interpreting Indira Gandhi's India', *Journal of Commonwealth and Comparative Studies*, 22: 128–47.

Mehta, M. (1994) *Rajiv Gandhi and Rama's Kingdom*. London: Yale University Press.

Mellor, J. W. (1966) *The economics of agricultural development*. Ithaca, NY: Cornell University Press.

Mellor, J. W. (1968) *Development in Rural India*. Ithaca, NY: Cornell University Press.

Mellor J. W. (1976) *The New Economics of Growth: A Strategy for India and the Developing World*. London, Cornell University Press.

Mendelsohn, O. (1978) 'Collapse of the National Indian Congress', *Pacific Affairs*, 51: 41–65.

Migdal, J. S. (1974) *Peasants, Politics and Revolution: Pressures Towards Political and Social Change in the Third World*. Princeton : Princeton University Press.

Migdal, J. S. (1988) *Strong Societies and Weak States: State-Society Relations and State Capabilities in the Third World*. Princeton, NJ: Princeton University Press.

Migdal, J. S. (2001) *State in Society: Studying How States and Societies Transform and Constitute One Another*. Cambridge: Cambridge University Press.

Mirchandani, G. (1977) *Subverting the Constitution*. New Delhi: Ajanta Books.

Mitra, A. (1977) *Terms of Trade and Class Relations*. New Delhi: Oxford University Press.

Mitra, S. K. (1992) *Power, Protest and Participation: Local Elites and the Politics of Development in India*. London: Routledge.

Mitra, S. K. (2005) *The Puzzle of India's Governance: Culture, Context and Comparative Theory*. London: Routledge.

Mitra, S. K. and Chiriyankandath, J. (1992) *Electoral Politics in India: A Changing Landscape*. New Delhi: Segment Books.

Mitra, S. K. and Lewis, R. A. (eds) (1995) *Subnational Movements in South Asia*. Boulder, CO: Westview Press.

Mitra, S. and Singh, V. B. (1999) *Democracy and Social Change in India: A Cross Sectional Analysis of the Indian Electorate*. New Delhi: Sage.

Mooji, J (2005) 'The politics of economic reform in India: a review of the literature', in C. Bates and S. Basu (eds), *Rethinking Indian Political Institutions*. London: Anthem Press, 1–20.

Morris-Jones, W. H. (1957) *Parliament in India*. London: Longmans.

Morris-Jones, W. H. (1967) 'The Indian Congress Party: dilemma of dominance', *Modern Asian Studies*, 1: 109–32.

Morris-Jones, W. H. (1971) *The Government and Politics of India*, 3rd edn. London: Hutchinson.

Morris-Jones, W. H. (1977) 'Creeping but uneasy authoritarianism', *Government and Opposition*, 12: 20–41.

Mouzelis, N. (1979) 'Modernisation, Underdevelopment, Uneven Development: Prospects for a Theory of Third World Formations', in *Journal of Peasant Studies*, 7(4): 190–227.

Mukhopadhya (2 March 2004) 'The taming of the shrew', *Hindustan Times*.

Narain, I. (1976) *Rural Elite in India*. New Delhi: Sage.

Narayan, Jayaprakash (1975) *Prison Diary*, edited with an introduction by A.B. Shah. [2nd and uncensored edition]. Bombay: Popular Prakashan, 1977.

Nayar, K. *The Judgement*, New Delhi: Vikas Books.

Noorani. A. G. (2000) *The RSS and the BJP: A Division of Labour*. New Delhi: LeftWord Books.

Nordlinger, E. (1981) *On the Autonomy of the Democratic State*. Cambridge, MA: Harvard University Press.

Nossiter, T. J. (1988) *Marxist State Governments in India: Politics, Economics and Society*. London: Pinter.

Nugent, N. (1990) *Rajiv Gandhi*. London: BBC Books.

Omvedt, G. (ed.) (1982) *Land, Caste and Politics in Indian Politics*. New Delhi: Left Books.

Omvedt, G. (1985) 'Capitalist agriculture and rural classes in India', in I. Khan (ed.), *Fresh Perspectives on India and Pakistan*. Oxford: Bougainvillea Books, 98–142.

Pal, C. (1984) *Centre-State Relations and Co-operate Federalism*. New Delhi: Sangam.

Pandey, G. (2006a) *Routine Violence: Nations, Fragments, Histories*. Stanford: Stanford University Press.

Pandey, G. (2006b) *The Construction of Communalism in Colonial North India*, 2nd edn. New Delhi: Oxford University Press.

Quigley, Declan (1993) *The Interpretation of Caste*. Oxford: Clarendon Press, viii, 184.

Quraishi, Z. M. (1973) *Struggle for Rashtrapati Bhawan*. New Delhi: Sterling Press.

Raj (1980) *Some Aspects of Wheat and Rice Policy in India, 1982*. Washington, DC: World Bank Publication.

Ranjit Sau (1977) 'Indian political economy 1967–1977: the marriage of wheat and whisky' *EPW*, 12(15): 615.

Rao A. and Rao, B. G. (1975) *Why Emergenc?* New Delhi: Parliamentary Press.

Ray, S. K., Chatterjee, V. and Venkatasubbiah (1982) *Political Development and Constitutional Change*. New Delhi: Sangam.

Robinson, M. and White, G. (eds) (1999) *The Democratic Developmental State*. Oxford: Oxford University Press.

Robinson, R. (1997) 'Weaving a Tale of Resistance', *EPW*, February 15–21, 32(7): 334–340.

Rudolph, L. and Rudolph, S. (1977) 'Backing into the future', *Foreign Affairs*, 55: 836–853.

Rudolph L. and Rudolph, S. (1978) 'To the brink and back: representative institutions and the Indian state', *Asian Survey*, 18: 379–400.

Rudolph, L. and Rudolph, S. (1987) *In Pursuit of Lakshmi: The Political Economy of the Indian State*. Chicago: University of Chicago Press.

Rudolph, S. (1965) 'Consensus and conflict in Indian politics', in *World Politics*, 13: 385–399.

Rueschemeyer, D. Stephens, E. H. and Stephens, J. D. (1992) *Capitalist Development and Democracy*. Cambridge: Polity Press.

Ruparelia, S. (2005) 'The temptations of presidentialism: an explanation of the evolving strategy of the BJP', in C. Bates and S. Basu (eds), *Rethinking Indian Political Institutions*. London: Anthem Press, 21–38.

Sahgal, N. (1982) *Indira Gandhi: The Road to Power*. New Delhi.

Sarkar, S. (1983) *Popular' Movements and 'Middle Class' Leadership in Late Colonial India: Perspectives and Problems of a 'History from Below'*. Calcutta: Bagchi for Centre for Studies in Social Sciences.

Sarkar, S. (1989) *Modern India: 1885–1947*, 2nd edn. Basingstoke: Macmillan.

Sathyamurthy, T. V. (November 1969) 'Crisis in the congress party', *World Today*, 25(11), 1127–1131.

Sathyamurthy, T. V. (1993) *Nationalism in the Contemporary World: Political and Sociological Perspectives*. London: Pinter.

Sathyamurthy, T. V. (ed.) (1996) *Class Formation and Political Transformation in Post-Colonial India*. New Delhi: Oxford University Press.

Savarkar, V. D. (1999) *Hindutva: Who Is a Hindu?* 7th edn. Mumbai: Swatantryaveer Savarkar Rashtriya Smarak.

Selbourne, D. (1977) *An Eye to India*. Harmondsworth: Penguin Books.

Selbourne, D. (1982) *Through the Indian Looking-Glass: Selected Articles on India, 1976–1980*. London: Zed Press.

Sethi, J. D. (1969) *India's Static Power Structure*. New Delhi: Sangam.

Sethi, J. D. (1974) *India in Crisis*. New Delhi: Sangam.

Shah, G. (1977) *Protest Movements in 2 Indian States*. New Delhi: Progress Publishing House.

Shanin, T. (ed.) (1987) *Peasants and Peasant Societies: Selected Readings*, 2nd edn. Oxford: Basil Blackwell.

Shourie, A. (1977) *Symptoms of Fascism*. New Delhi: Vikas.

Shourie, A. (1982) *Mrs. Gandhi's Second Term*. New Delhi: Vikas.

Shourie, A. (1997) *A Secular Agenda for Saving Our Country, for Welding It*. New Delhi: HarperCollins.

Simpson, A. W. B. (2001) *Human Rights and the End of Empire: Britain and the Genesis of the European Convention*. Oxford: Oxford University Press.

Singh, G. (2003) in Burnell, P. (ed.) *Democratization Through the Looking Glass*, 216–230.

Singh, K. (1991) *A History of the Sikhs. Volume Two: 1839–1988*. Bombay: Oxford University Press.

Singh, K. A. (1991) *A History of the Sikhs. Volume One: 1469–1839*. Bombay: Oxford University Press.

Siwach, J. R. (1979) *The Politics of President's Rule in India*. Simla: Sterling Press.

Skocpol, T. (1979) *States and Social Revolutions: A Comparative Analysis of France, Russia, and China*. Cambridge: Cambridge University Press.

Skocpol, T. (1982) 'Bringing the state back in', *Items* (SSRC Bulletin USA) 36: 23–53.

Smith, D. E. (1968) *India as a Secular State*. Princeton: Princeton University Press.

Somjee, A. H. (1979) *The Democratic Process in a Developing Society*. London: Macmillan.

Srinivas, M. N. (1987) *The Dominant Caste and Other Essays*. New Delhi: Oxford University Press.

Srinivas, M. N. (1995). *Social Change in Modern India*. Hyderabad: Orient Longman.

Stokes, E. (1978) *The Peasant and the Raj*. Cambridge: Cambridge University Press.

Tambiah, S. J. (1998) 'The crisis of secularism in India' in R. Bhargava (ed.), *Secularism and Its Critics*. New Delhi: Oxford University Press, 418–53.

Tarlo, E. (2003) *Unsettling Memories: Narratives of India's 'Emergency'*. New Delhi: Permanent Black.

Thakur, C. P. (Chandreshwar Prasad), 1999-: India under Atal Behari Vajpayee: the BJP era /C.P. Thakur, Devendra P. Sharma. New Delhi: UBS Publishers.

Thakur, R. (1995) *The Government and Politics of India*. Basingstoke: Macmillan.

Tharoor, S. (1989) *The Great Indian Novel*. London: Penguin

Tomlinson, T. (1993) *The Economy of Modern India*. Cambridge: Cambridge University Press.

Torri, F. (1975) 'Factional politics and economic policy: the case of India's bank nationalisation', *Asian Survey*, 15: 1077–96.

Toye, J. F. J. (1976) 'Economic policies in India during the emergency', *World Development*, 5: 303–20.

Toye, J. F. J. (1979) *Public Expenditure and Indian Development Policy*. Cambridge: Cambridge University Press.

Toye, J. F. J. (1981) *Dilemmas of Development: Reflections on the Counter-Revolution in Development Economics*, 2nd edn. Oxford: Blackwell.

Tully, M. and Satish Jacob (1985) *Amritsar: Mrs Gandhi's Last Battle*. Calcutta: Rupra paperbacks.

Vanaik, A. (1990) *The Painful Transition: Bourgeois Democracy in India*. London: Verso.

Vanaik, A. (1997) *The Furies of Indian Communalism: Religion, Modernity, and Secularization*. London: Verso.

Varshney, A. (1993) 'Battling a past, forging a future: Ayodhya and beyond', in M. Boulton and P. Oldenburg (eds), *India Briefing 1993*. Boulder, CO: Westview Press.

Varshney, A. (2002) *Ethnic Conflict and Civic Life: Hindus and Muslims in India*. New Haven: Yale.

Veer, Peter van der (1994) *Religious Nationalism: Hindus and Muslims in India*. New Delhi: Oxford University Press.

Venkatasubbiah, H. (1969) *Anatomy of Indian Planning*. London: Institute of Pacific Relations.

Wadhwa, K. K. (1975) *Minority Safeguards in India: Constitutional Provisions and their Implementation*. New Delhi: Vikas.

Washbrook, D. (1976) *Emergence of Provincial Politics: Politics in the Madras Presidency 1870–1920*. Cambridge: Cambridge University Press.

Weiner, M. (1978) *Sons of the Soil: Migration and Ethnic Conflict in India*. Princeton: Princeton University Press.

Weiner, M. (1982) 'Congress restored: continuities and discontinuities in Indian politics', *Asian Survey*, 22: 339–55.

Weiner, M. and Katzenstein, M. F. (1981) *India's Preferential Policies: Migrants, the Middle Classes and Ethnic Equality*. Bombay: Oxford University Press.

Weiss L. and Hobson, J. H. (1995) *States and Economic Development: A Comparative Historical Analysis*. Cambridge: Polity Press.

Wood, J. R. (1975) 'Extra-Parliamentary Opposition in India', in *Pacific Affairs*, 48(3): 313–34.

Wood, J. R. (1984a) 'Continuity and crisis in Indian state politics', in John R. Wood (ed.) *State Politics in Contemporary India: Crisis or Continuity?* Boulder, West View Press, 1–20.

Wood, J. R. (1984b) 'Congress restored? The "KHAM" strategy and Congress I recruitment in Guajarat', in John R. Wood (eds) *State Politics in Contemporary India: Crisis or Continuity?* Boulder: Westview, 197–228.

Zavos, J. (2000) *The Emergence of Hindu Nationalism in India*. New Delhi: Oxford University Press.

Zavos, J., Wyatt, A. and Hewitt, V. (eds) (2004) *The Politics of Cultural Mobilization in India*. New Delhi: Oxford University Press.

Index

20 point programme 132, 145

accountability, lack of 141
adivasis 73
Advani, L.K. 184
Advisory Committee on Minority
  Rights 52
affirmative action, and religious
  sensitivities 51–2
Agenda for National Government 193
Agricultural Prices Commission
  (APC) 72
agriculture 69–73, 102–3, 167
Allahabad verdict 118, 119, 126
All India Congress Committee
  (AICC) 44, 64, 65, 82, 89
Ambedkar, Dr B.R. 56–7
analyses of Emergency:
  Bonapartism 152; Marxist 25–7,
  28; from point of view of
  beneficiaries 19–20;
  sociological-historical 21–2;
  systemic liberal 17–19, 27–8
authoritarianism 150
authoritarian populism 27, 169
authority, constitution of 203
Ayodhya 176–8, 181, 182, 183, 185–8

Babri Masjid Action Committee 177
Bahuguna, H.N. 137, 147
Bajrang Dal 163, 170, 185, 187
Balwantray Mehta Report 71
Bano, Shah 174–5
Bardhan, P. 25–6
Barooah, D.K. 132, 146
Basu, Durga Das 59–60
Bayly, S. 68
Belchi massacre 156
Bhindranwale, Sant 168–9
Bihar 105–7, 109, 153, 165

bill of rights 49–50, 58
BJP: coalition government 190, 192;
  coalition politics 195; corruption 202;
  disagreements with RSS 193; electoral
  success 185; explanation of rise 31;
  factors contributing to rise 196–7;
  gains in 1989 election 178–9;
  ideological coherence 194; and
  Janata Party 154–5; links with
  RSS 190–1; poor performance
  in 1984–5 elections 171; pre-election
  alliances 191; relationship with
  RSS 163
BJS 47–8, 73–4, 91, 101, 107, 154–5
BKD 91
Bonaparte, Louis 23–4
bourgeois democracy 23
bullock capitalism 22
bureaucratic efficiency 128
bureaucratic elite 30

cabinet secretariat 96
castes: approach of post-independence
  parties 46; defining role in
  politics 199; and economic
  power 68–9; and political power
  75–6; as sources of resistance 204;
  unrest 183; violence 156
censorship 123, 127, 148
centralisation: 42nd amendment 138;
  authoritarianism 150;
  counterproductive 141–2; fears of 57,
  61–2; following 1969 split 96; as
  ideology 166; under Indira Gandhi
  18–19, 84–5; post independence 53–4;
  under State of Emergency 126–7
centrality 14
Chad, Krishan 141
Chatterjee, P. 34–6, 149, 205
Chavan, Y.B. 101

Chief Justice 98, 112
chief ministers 56, 80
civil rights 201
civil society 32, 34
class 28, 204
class alliance 22
The Class Struggle in France 23
coalition government 90, 180, 199
Code of Criminal Procedure,
    section 125 174–5
collaboration 21, 22
colonialism 19, 43
commodity prices 103
communalism 174
Communist Party of India (CPI) 15,
    88, 100, 119, 130, 135, 143, 145, 151
Community Development
    Programme 65, 70
complexity, political 202
conflict, intragovernmental 64
Congress: 1978 split 165; accusations
    of fascism 112; courting middle
    class 129; crisis of legitimacy 106;
    electoral defeat 148–9, 152;
    elimination in 1989 and 1990
    elections 179; expulsions 137;
    factional alignments 143; Hindu
    majoritarianism 178; loss of support
    in Gujarat 115; national identity 138;
    overt meeting of rebels 119;
    relationships with castes 164;
    restructuring after split 97; seen as
    appeasing Muslims 175
Congress Election Committee
    (CEC) 44, 92
Congress for Democracy 148
Congress Forum for Socialist
    Action 15, 82, 90
Congress-I 101
Congress-O 87, 89, 90–1
Congress Parliamentary Board
    (CPB) 44
Congress Party: aggregation 66;
    basis of success 18; collapse of
    state-based systems 74;
    determinants of collapse 196;
    development 42–3; economic
    policies 83; electoral weakness 84;
    as elite 42; flexibility 43–4, 64;
    ideological disputes 83; internal
    conflict after independence 45;
    internal tensions 43; loss of support
    76; loss of votes 1967 election 78–9;
    relationship between government
    and organisation 64–5; relationship

with state 46; reliance on elites 67;
    restructuring after split 94; split,
    1969 15, 18, 19, 87–8, 93–4, 96;
    state-society model 66; structure 44;
    survival in south 152; weakened by
    divisions 73
Congress-R 87–90, 91, 164
Congress Working Committee
    (CWC) 44–5, 147
Congress Working Committee-R
    (CWC-R) 91–2
Conservation of Foreign Exchange
    and Prevention of Smuggling
    Act 124, 126
Constituent Assembly 54–8, 60
constitution 48–51; 42nd amendment
    134–6, 142, 145; 44th amendment 158;
    52nd amendment 173; amendments
    under state of emergency 125–6,
    134; Article 356 200; debates on
    change 134; First Amendment 70;
    flexibility 60–1; Rajiv's commitment
    to amend 173; reactions to 61;
    relationship between governors
    and chief ministers 80; restoration
    of credibility 158
conversions, to Islam 170–1
Corbridge, S. 29–31, 33
corruption 94–5, 172, 201, 202
courts 98, 200–1
court system 57–8
credibility 152
crisis, 1969 15

dalits 52, 73, 155, 156, 201
decentralisation 71
Defence and Internal Security of
    India Act, 1975 123
Defence of India Rules 110–11
deinstitutionalisation debates 74, 203
Delhi, detentions 122–3
Delhi Development Authority,
    intimidation 139
Delhi Municipal Corporation 139
democracy 23, 75
democratic system 62
democratisation 35
Desai, Morarji 76–7, 85, 113, 155, 160, 161
detentions 121–3, 148
Dharia, Mohan 110, 118
dictatorship, suggestions of 111
district Congress Committees 132
dominant coalition 25–6, 30
double clearing 97
dysfunctionality, of government 127

economic crisis 100
economic policies 83
economic power, and social
  status 68–9
economic programmes, under State
  of Emergency 128
economic recovery 129
economy 110–11, 130
education, and secularism 161–2
*The Eighteenth Brumaire of Louis
  Bonaparte* 23–4
elections: 1967 78–9; 1971 15, 92; 1972
  92, 93–4; 1984–5 171–2; 1989 178–9;
  1990 179; 1991 184–5; 1993 188;
  1996 189–90; 1998 191–2; 2004 199;
  following Emergency 145–7;
  Gujarat 113–14; relaxation of
  detention and censorship 148;
  voting patterns 164–5, 179
electoral instability 160
elites 24–5, 67, 151
embedded autonomy 37
emergency interim budget, 1974 104
Engels, Friedrich 23–5
equality 49
executive 54
executive democracy 21
External Emergency 112, 113, 122,
  124, 157

family planning 140
farmers' movements 156
farm lobbies 103
federalism 53
Finance Commission 54
financial institutions, nationalisation 85
fixers 97
foreign industry 130
Foucault, Michel 14
fragmentation 140–1, 156
Frankel, F. 22, 68–9, 75–6, 99, 151
fundamental rights 48, 58–9, 70, 98, 141

Gandhi, Indira: appeal to Supreme
  Court 117; assassination 171;
  authoritarianism 166; battle with
  syndicate 82; centralisation 18–19;
  choice as prime minister 77;
  consolidation of support 126; decision
  not to resign 116–17; declarations
  of support 117–18; electoral
  malpractice 16, 95, 112, 115;
  as element of coalition 99–100;
  encouragement of political defections
  88; ideology 97; increased power of

prime minister 96; interventionism
  86; justification of state of
  Emergency 17; manipulation of
  media 158; move away from
  secularism 168; opposition to
  Sanjiva Reddy 84–6; as political
  manipulator 20; political rebirth
  156–7; populism 99; presidential
  ordinances 100; pressure to resign
  115–16; re-election to parliament 159;
  resignation 148; return as prime
  minister 164; secularism 78; style
  of leadership 17–18; support for
  V.V. Giri 85–6; vision of India 152
Gandhi, Mohandas K. 42
Gandhi, Rajiv 38, 172–3, 184
Gandhi, Sanjay 95–6, 138–41, 144, 167
garibi hatao 2, 15, 38, 91, 99, 118,
  151, 197
Ghafoor, Abdul 106
Giri, V.V. 85–6
Gokhale, H.R. 133
government 54–5, 140–1, 145
governmentality 34
Government of India Act, 1935 59
governors 56–7, 80
grain markets 129
grain prices 104
Grand Alliance 91
Gujarat 16, 104–5, 106, 107, 113–15

Hanson, A.H. 71
harijans 52, 90, 155, 170
Harriss, J. 29–31, 33
Hindu Code Bill 51
Hinduism 68, 73
Hindu Mahasabha 41
Hindu majoritarianism 30–1, 178
Hindu nationalism: analytical
  writing 32–9; approach to social
  and political change 45; attempts
  to exclude 161; distinct from Indian
  nationalism 40–2; fear of 169;
  fear of rise 193; as reclaiming
  nation 177; response to treatment of
  Muslims 175–6; role of elites 41–2
Hindutva: adoption by middle
  classes 185; analytical writing 32–9;
  challenge to state power 30–1;
  compared to European fascism 46;
  contributory factors 196; cultural
  nationalism 46; elitism 198; as
  ideological identity 41; mobilisation
  of social support 14; rise of 164;
  symbolism 188–9; as threat 204

Hobson, J.H. 37
Hussain, Dr Zakir 83–4

identities 43, 53, 67
ideology 75, 151, 202
inclusion, educational and
  cultural 34–5
INC Report on The Fourth General
  Election 82
Indian nationalism, distinct from
  Hindu nationalism 40–2
Indian scholarship, influences on 14
indirect rule 35
Indo-Pakistan War 1971 100–1
Industrial Disputes (Amendment)
  Act 130
industry, corruption 102
inflation 128–9
instrumentalism 20
Intensive Agricultural District
  Programme 71
Islam, mass conversions to 170–1

Jaffrelot, C.: Ayodhya 186; BJS 47–8;
  Hindu nationalism 41–2; 'impossible
  assimilation' 102; populism 99; Rath
  Yatra 184; secular hegemony 38;
  secular ideology 74
Janata Dal 178–9, 183, 184–5, 189
Janata Party 153–5, 157–60, 162
Jan Morcha 178
jati 43, 52, 68–9
Joshi, S.S. 194
JP Movement 16, 94–5, 102, 107–9, 119

Kamaraj Plan 76
Kargil 192
kar sevaks 184
Kesavananda Bharati verdict 98
Kochanek, S. 19, 66
Kohli, A. 26–7
Kothari, R. 20–1, 149, 203

land-ceiling Acts 70, 131–2
land reform 69–73, 131–2
legal aid 200
legislation, lists of 53–4
legitimation, of state 29
Leys, C. 25
liberalisation, under National
  Democratic Alliance (NDA) 192
liberalism 201
local governments, collapse 81
Lok Sabha 54, 90, 125, 146, 199, 201

Maintenance of Internal Security Act
  (MISA) 100, 110–11, 123, 124, 140–2
Mandal Commission 31, 183
Manifesto of the Communist Party 23
Manor, J. 18
Marx, Karl 23–5
Marxism 14, 23
media 123
Meenakshipuram 170–1
Members of the Legislative Assembly
  (MLA) 56
Migdal, J.S. 29
Minorities Commission 157
minority government 199
Mishra, L.N. 95, 111–12
mobilisation: caste and class 181;
  ethno-religious 163, 187–8;
  horizontal 75; under Janata 153;
  of political elite 42; social and
  cultural movements 72–3; of social
  support 36–9
modernisation, context 53
monsoon, 1972 103
Morris-Jones, W.H. 17–18
movements, social and cultural 72
Muslims, demonstrations 175
The Muslim Women (Protection of
  Rights on Divorce) Bill 175–6

Narain, I. 71
Narain, Raj 16
Narayan, J.P. 107–8, 110, 119, 148
National Democratic Alliance
  (NDA) 182, 192–3, 199
National Extension Service 70
National Front 180
nationalisation 85, 152
Nehru, J. 42, 45, 51, 66–7, 76
Nehruvian socialism 33, 197
neo-Gandhianism 154
neo-statism 37
networks 94, 97, 174
NGOs 167–8
Nijalingappa, S. 83–4, 86
no-confidence motions 95–6
northern India 81, 155–6

OBCs 10, 31, 52, 73, 179, 183, 188, 199
opposition, detention 121–3
opposition parties 62–3, 114, 148, 166,
  198; *see also* individual parties
opposition rally 119, 120
ordinances, power of president 55–6
other backward classes *see* OBCs

panchayat raj 70–1
Parliament, power of 64, 98
Patel, Chimanbhai 105–7
Patel, Sardar 41, 45
patron-client systems 65, 66
PCCs 65
peasant groups, political
allegiance 152–3
personalisation, of power 127
Planning Commission 71
political complexity 202
*The Political Economy of Development
in India* 26
political instability 81–2
political patronage 75
political society, distinct from civil 34
political system, reinstitutionalisation
under Rajiv 173
political tension 112
politics 20, 203
populism 20
post-modernism 33
post-structuralism 33
poverty 52
power, social 67
power relations, in government 64
president, role and powers 54–6, 200
presidential elections 86–7
presidential ordinances 100
President's Rule 57, 79–80, 88, 166
press, censorship 123, 127
Press Council Amendment
Ordinance 124
Prevention of the Publication
of Objectionable Matter
Ordinance 123–4
preventive detention 59, 100–1
Preventive Detention Act, 1950 59
prices, attempts to stabilise 103–4
prime minister 54–6, 76, 96
prime minister's secretariat 64
private companies, restructuring 130
procedure, disputes over 80
processions, as confrontation 183
property rights 201
protective discrimination 52–3
Punjab 168–9
Punjab Accord 173

railway strike 111
Rajasthan 80
Rajya Sabha 54–5
rallies 17
Ram, Jagjivan 118, 143, 147–8

Rama Rao, N.T. 166
Ramjanmabhoomi agitations 181
Rao, Narasimha 184–5
Rath Yatra 183–4
rationing 104
Ray, S.S. 120–1
Reddy, Brahmananda 122
Reddy, Sanjiva 83–6, 149
redistributive policies 72
reductionism 202
reflation 129
regionalism 19
Reinventing India 29–31
religion 67–8, 162
religiosity 162–3
religious freedom 50, 51
religious nationalism 181
religious pluralism 67
religious sensitivity 51–2
resistance, class and caste based 204
revisionism, understanding of state 32
riots 106–7, 171
RSS 33; attitude to state 46;
banning 187; blamed for civil
unrest 107; compared to European
fascism 47; illegitimacy in
government 190–1; kudos 154;
normalisation 191; relationship
with BJP 163; repression 101; rise
of 42; role within BJS 73–4; as
social network 47; as source of
ideology 132; tensions with BJS 91;
vision of India 61
Rudolph, L. and S. 21–2, 152
Rueschemeyer, D. 37

Samachar (news agency) 124
sangathan 46
Sangh Parivar 11, 163, 181, 182, 185,
190–1, 193, 198, 199
Sarkaria Commission 166
Savarkar, V.D. 42
scandal 95
secularism 33, 48, 161–2, 169, 181–2,
189, 192, 197, 204
Sethi, J.D. 20
Shah Commission 139, 145, 153, 158
Shariat law 174–5
Shastri, Lal Bahadur 76–7
Shekhar, Chandra 184
Shukla, V.C. 123
Sikh extremism 168–9
Simon Committee of the Indian
Constitution 49

Singh, Charan 86–7, 158–9, 160
Singh, Swaran 118
Singh, V.P. 178–80, 183, 184
Skocpol, T. 28–9
slum clearance 139–40
Smith, D.E. 52
social exclusion 167
social identity 32
social power 67, 68–9
social reform 131
social status, and economic power 68
social welfare, opposition
    support 114–15
societal autonomy 36
special powers legislation 100
state: assemblies 56; corporatism 21–2;
    decline 38; depoliticisation 173;
    elections 15; governments 80,
    157–8, 186–7, 201; as instrument of
    change 43; liberal 22; as multiple
    elements 195; relation to dominant
    groups 68–9
The State and Poverty in India 26
State of Emergency 59–60
State of Internal Emergency: attempts
    to link with populism 145;
    background 15–17; chaos 150;
    collapse 146; confusion 127–8;
    consolidation 137–8; constitutional
    amendments 125–6; drafting of
    letter to President 121; economic
    programmes 128; effects of
    collapse 196; events immediately prior
    to declaration 119–20; extent of Prime
    Minister's power 121–2; government
    defence of 128; ideological
    inconsistency 151; institutional
    changes to central government 126–7;
    institutionalisation 142; lack of
    protest 124; legal foundations for
    reform 133; legitimation 150–1;
    normalisation process 124–5;
    notification of cabinet members 122;
    public comment by key figures 133;
    social reform 131; use of coercive
    measures 130; wording of
    proclamation 121
states 65–6, 103, 167
States and Social Revolutions 28–9
state-society relations: changes 195;
    changing political landscape 193;

as contested area 28; continuum
    197–8; effect of 1969 split 93;
    emphasis on 203; reconfiguration
    204; redefinition 151
sterilisation campaign 140
strikes, 1974 104
subaltern classes 34–5, 198–9
subaltern scholarship 42–3
subaltern studies 32
Sukhadia, Mohan Lal 80
Supplementary Report on
    Fundamental Rights 50
Supreme Court 57, 133, 200
swadeshi 188
Swaran Singh committee
    reports 134–5
syndicate 81–2

Tamil Nadu 137
Tarlo, Emma 14
technology, in government 173
Thapar, R. 149
theories of state 28, 32
third world 28
Turkman Gate killings 139–40

unemployment 130
Union Constitution Committee 53
Union Powers Committee 53
Uttar Pradesh 79, 86–7, 137, 153,
    184, 185

Vajpayee, A.B. 107, 116, 154, 160,
    190, 192–3
Veer, Peter van der 32
VHP 163, 170, 185
VHP-BJP processions 183
village uplift programme 70
violence 139, 156, 184, 185
voting, and representation 63

Weiss, L. 37
welfare 203
West Bengal 80
wheat trade 103
wholesale price index 142

Young Turks 83–4
Youth Congress 138–40, 143–5

Zavos, J. 202

Printed in the United Kingdom by
Lightning Source UK Ltd., Milton Keynes
139802UK00001B/51/P